The Wonder Book of Geometry

DAVID ACHESON

THE WONDER BOOK OF GEOMETRY

a mathematical story

OXFORD
UNIVERSITY PRESS

OXFORD
UNIVERSITY PRESS

Great Clarendon Street, Oxford, OX2 6DP,
United Kingdom

Oxford University Press is a department of the University of Oxford.
It furthers the University's objective of excellence in research, scholarship,
and education by publishing worldwide. Oxford is a registered trade mark of
Oxford University Press in the UK and in certain other countries

First Edition published in 2020

Impression: 3

Published in the United States of America by Oxford University Press
198 Madison Avenue, New York, NY 10016, United States of America

British Library Cataloguing in Publication Data

Data available

Library of Congress Control Number: 2020932235

ISBN 978–0–19–884638–3

Printed and bound in Great Britain by
Clays Ltd, Elcograf S.p.A.

Contents

1. Introduction   1

2. Getting Started   4

3. Euclid's *Elements*   9

     Euclid, 1732   12

4. Thales' Theorem   14

     The Mathematical World of Ancient Greece   18

5. Geometry in Action   20

6. Pythagoras' Theorem   26

7. 'In Love with Geometry'?   36

     371 Proofs of Pythagoras   42

8. 'Imagine my Exultation, Watson…'   44

9. Congruence and Similarity   50

     The Golden Ratio   58

10. Conversely…   60

11. Circle Theorems   68

12. Off at a Tangent   73

13. From Tangents to Supersonic Flow   79

     Galileo and Thales' Theorem   84

14. What is π, Exactly?   86

15. The Story of the Ellipse   94

16. Geometry by Coordinates   101

        Inspector Euclid Investigates...   106

17. Geometry and Calculus   108

18. A Royal Road to Geometry?   114

19. Unexpected Meetings   122

20. Ceva's Theorem   129

        Some Further Slices of Pi   136

21. A Kind of Symmetry   138

22. 'Pyracy' in Woolwich?   145

23. Fermat's Problem   154

24. A Soap Solution   164

25. Geometry in *The Ladies' Diary*   171

        Euclid, 1847   178

26. What Euclid Did   180

27. Euclid on Parallel Lines   189

        Proof by Picture?   196

28. 'A New Theory of Parallels'?   198

29. Anti-Euclid?   205

30. When Geometry Goes Wrong...   213

31. New Angles on Geometry   223

32. And Finally...   231

Notes   241

Further Reading   265

Acknowledgements   269

Publisher's Acknowledgements   270

Picture Credits   271

Index   273

Introduction

It all started at school, one cold winter morning in 1956, when I was ten.

Mr. Harding had been doing some maths at the blackboard, with chalk dust raining down everywhere, when he suddenly whirled round and told us all to draw a semicircle, with diameter AB.

Then we had to choose some point P on the semicircle, join it to A and B by straight lines, and measure the angle at P (Fig. 1).

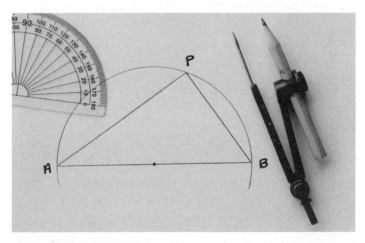

Fig. 1 Thales' theorem.

I duly got on with all this, casually assuming that the angle at P would depend on where P is, exactly, on the semicircle.

But it *doesn't*.

It's always 90°.

* * *

At the time, I had no idea that mathematics is full of surprises like this.

I had no idea, either, that this is one of the first great theorems of geometry, due to a mathematician called Thales, in ancient Greece. And according to Thales – so it is said – the key question is always not 'What do we know?' but rather '*How do we know it?*'

Why is it, then, that the angle in a semicircle is always 90°?

The short answer is that we can *prove* it, by a sequence of simple logical steps, from a few apparently obvious starting assumptions.

"YOU WANT PROOF? I'LL GIVE YOU PROOF!"

Fig. 2 The importance of proof.

And by doing just that, in the next few pages, I hope to not only lay some foundations for geometry, but do something far more ambitious.

For, with geometry, it is possible to see something of the whole nature and spirit of mathematics at its best, at almost any age, *within just half an hour of starting*.

And in case you don't quite believe me…

Getting Started

The first really major idea is that of *parallel lines*.

These are lines, in the same plane, which never meet, no matter how far they are extended.

And I will make two assumptions about them.

Parallel lines

Imagine, if you will, two lines crossed by a third line, producing the so-called *corresponding angles* of Fig. 3.

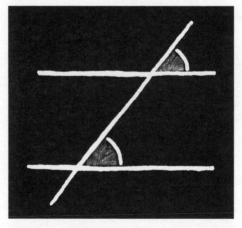

Fig. 3 Corresponding angles.

Then, throughout most of this book, I will assume that

(1) If two lines are parallel, the corresponding angles are equal.

(2) If corresponding angles are equal, the two lines are parallel.

These assumptions are rooted in the intuitive notion that parallel lines must be, so to speak, 'in the same direction', but however obvious (1) and (2) may seem, they *are* assumptions.

And, even at this early stage, it is worth noting that they amount to two very different statements.

In effect, (1) helps us use parallel lines, while (2) helps us show that we have some.

Angles

We will measure angles in *degrees*, denoted by °, and the two parts of a straight line through some point P form an angle of 180° (Fig. 4).

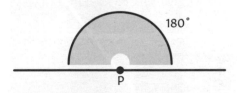

Fig. 4 A straight line.

A *right angle* is half this, i.e. an angle of 90°, and the two

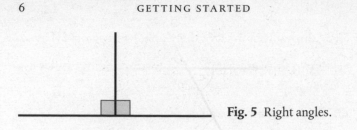

Fig. 5 Right angles.

lines forming it are then said to be perpendicular (Fig. 5).

Opposite angles

When two straight lines intersect, the so-called *opposite angles* are equal (Fig. 6).

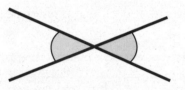

Fig. 6 Opposite angles.

Alternate angles

If two lines are parallel, and crossed by a third line, then the so-called *alternate angles* are equal (Fig. 7).

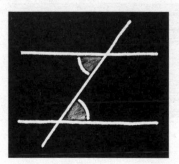

Fig. 7 Alternate angles.

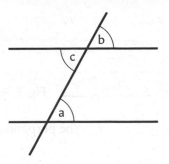

Fig. 8 Proof that alternate angles are equal.

This is because, in Fig. 8, $a = b$ (corresponding angles) *and* $b = c$ (opposite angles). So $a = c$.

The argument works 'in reverse', too, so that if alternate angles are equal, the two lines must be parallel.

And with these ideas in place, we are now ready to prove the first theorem which, in my view, is not obvious at all...

The angle-sum of a triangle

The three angles in any triangle add up to 180° (Fig. 9).

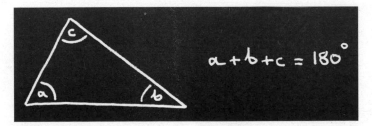

Fig. 9 Angles in a triangle.

To prove this, draw a straight line through one corner, parallel to the opposite side (Fig. 10).

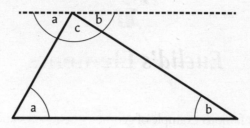

Fig. 10 Proof of the angle-sum of a triangle.

The angles *a* are then equal (alternate angles).

The angles *b* are also equal, for the same reason.

Finally, the new line is straight, so $a + b + c = 180°$, which completes the proof.

Euclid's Elements

The most famous example of geometry being presented in this concise, deductive, and carefully ordered way is the *Elements*, written by Euclid of Alexandria (Fig. 11), in about 300 BC.

Fig. 11 Euclid.

It is best to be clear from the outset, I think, that the precise theorems and proofs of Euclid's *Elements* (Fig. 12) are essentially about imaginary objects.

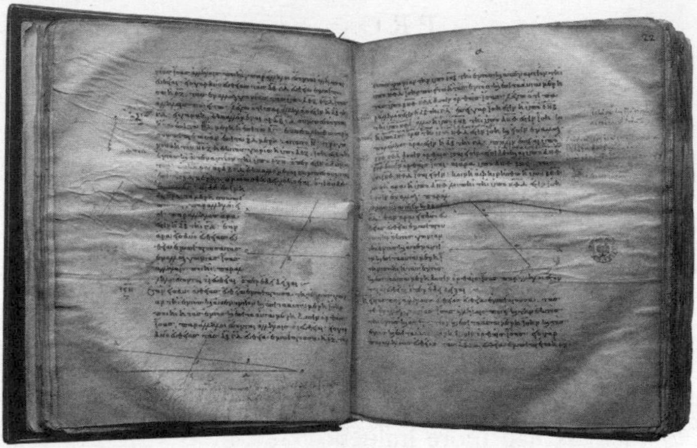

Fig. 12 The oldest surviving copy of Euclid's *Elements*, MS D'Orville 301, copied by Stephen the Clerk for Arethas of Patras, in Constantinople in AD 888.

A Euclidean straight line, for instance, isn't just 'perfectly' straight—it has *zero thickness*. So even if I could draw one properly, you wouldn't be able to see it.

And a point isn't a blob of small dimension—it has no dimension at all. Or, as Euclid put it:

A point is that which has no part.

It should be said, too, that Euclid makes no use of what we would call 'measurement units' for length. And there are no degrees in Euclid; the nearest he comes to having a unit for angle is the concept of *right angle*, which he uses a great deal (Fig. 13).

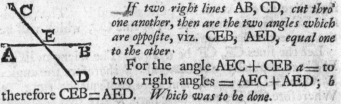

Fig. 13 Proof that opposite angles are equal, from a 1732 edition of Euclid's *Elements*.

In spite of this, and the austere style of exposition, the *Elements* has had more influence, and more editions, than almost any other book in human history.

In the end, however, there can be no single 'best' way of doing geometry, and we all have to find our own path into the subject.

And if, in this book, I unashamedly assume more than Euclid does, it is because I want to proceed more quickly to interesting and surprising results…

Euclid, 1732

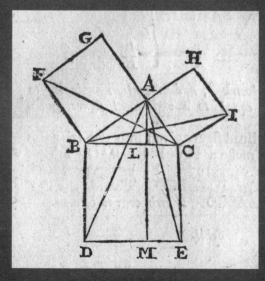

One of the most popular early editions of Euclid was by **Isaac Barrow.** It was first published in 1660, but my own copy dates from 1732.

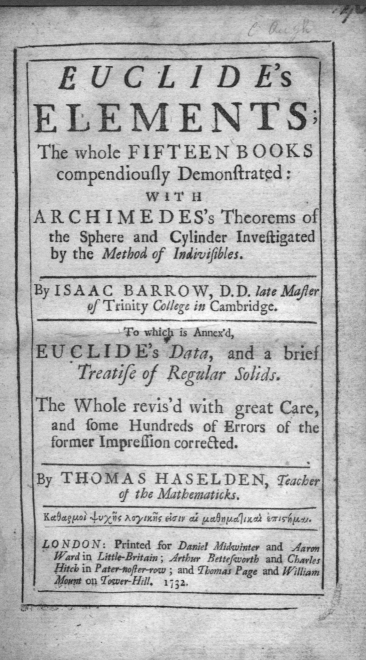

EUCLIDE's ELEMENTS;

The whole FIFTEEN BOOKS
compendiously Demonftrated:

WITH

ARCHIMEDES's Theorems of
the Sphere and Cylinder Inveftigated
by the *Method of Indivifibles.*

By ISAAC BARROW, D.D. *late Mafter
of* Trinity *College in* Cambridge.

To which is Annex'd,

EUCLIDE's *Data*, and a brief
Treatife of Regular Solids.

The Whole revis'd with great Care,
and fome Hundreds of Errors of the
former Impreffion corrected.

By THOMAS HASELDEN, *Teacher
of the Mathematicks.*

Καθαρμοὶ ψυχῆς λογικῆς εἰσιν αἱ μαθηματικαὶ ἐπιςῆμαι.

LONDON: Printed for *Daniel Midwinter* and *Aaron
Ward* in *Little-Britain*; *Arthur Bettefworth* and *Charles
Hitch* in *Pater-nofter-row*; and *Thomas* Page and *William
Mount* on *Tower-Hill.* 1732.

Thales' Theorem

Thales' theorem says that the angle in a semicircle is always 90°.

And, to prove it, we need just one or two more key ideas.

Congruent triangles

Congruent triangles are ones which have *exactly the same size and shape*.

And the most obvious way of fixing the exact size and shape of a triangle is, perhaps, to specify the lengths of two sides and the angle between them.

This leads to a very simple test for congruence, known informally as 'side-angle-side', or SAS (Fig. 14).

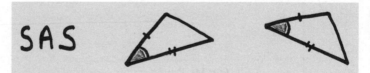

Fig. 14 Congruence by SAS.

Isosceles triangles

An isosceles triangle is one in which two sides are equal.

Triangles of this kind play a major part in geometry, largely because *the 'base' angles of an isosceles triangle are equal* (Fig. 15).

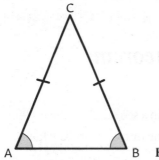

Fig. 15 An isosceles triangle.

Many people, I think, find this particular result rather obvious. After all, if we 'nip round the back' of an isosceles triangle it will look exactly the same.

A more formal way of proving the result is to introduce the line CD bisecting the angle at C (Fig. 16).

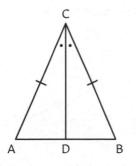

Fig. 16 Proof that the base angles of an isosceles triangle are equal.

The triangles ACD and BCD are then congruent by SAS, and one is, in fact, a 'mirror image' or 'overturned' version of the other. In particular, then, the angles at A and B must be equal.

(If *all three* sides of a triangle happen to be equal it is said to be *equilateral*. The triangle is then isosceles in three different ways, so all three of its angles are equal.)

Circles

The defining property of a circle is that all its points are the same distance from one particular point, called the *centre*, O.

Some other common terminology is introduced in Fig. 17.

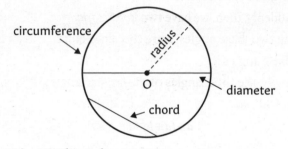

Fig. 17 The circle.

And this gives us all we need to prove Thales' theorem.

Thales' theorem

We want to prove that if P is any point on the semicircle in Fig. 18, then $\angle APB = 90°$, where $\angle APB$ denotes the angle between AP and PB.

Now, the simplest way of using the fact that P lies on the semicircle, surely, is to draw in the line OP and observe that

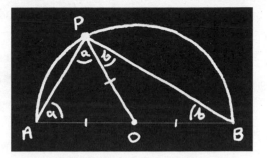

Fig. 18 Proof of Thales' theorem.

OP = OA = OB, because all points on a circle are the same distance from its centre.

Suddenly, then, we have two *isosceles triangles*, AOP and BOP.

The two 'base angles' a are therefore equal, and so are the two base angles b.

Finally, the three angles of the large triangle APB must add up to 180°, so

$$a+(a+b)+b=180°$$

and therefore $a + b = 90°$. In consequence, $\angle APB = 90°$, which proves the theorem.

And in all the years since I first saw this proof, on a cold winter morning in 1956, I have never forgotten it.

After all, the result is, at first sight, rather difficult to believe, yet just a few minutes later we find ourselves saying, arguably: 'Oh, it's sort of obvious, really, isn't it—*when you look at it the right way.*'

And in my experience, at least, this is often one of the hallmarks of mathematics at its best.

the mathematical world of ANCIENT GREECE

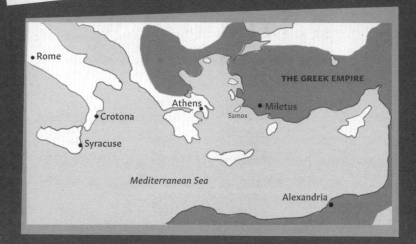

Thales lived in Miletus.

Pythagoras came from the island of Samos, but later moved to Crotona.

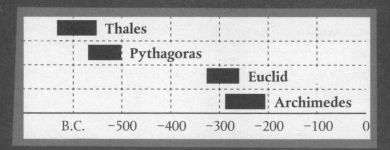

Plato's Academy in Athens had this famous inscription over its entrance:

ΑΓΕΩΜΕΤΡΗΤΟΣ ΜΗΔΕΙΣ ΕΙΣΙΤΩ

"Let no one ignorant of geometry enter here"

The Pharos Lightouse, Alexandria

Euclid wrote *The Elements* in Alexandria.

Archimedes lived and worked in Syracuse.

Geometry in Action

Throughout history there have been practical applications of geometry, and one of the earliest was Thales' attempt to calculate the height of the Great Pyramid in Egypt.

Thales and similar triangles

Thales measured the shadow of the Great Pyramid cast by the Sun, and by adding half the pyramid's base determined the distance L in Fig. 19.

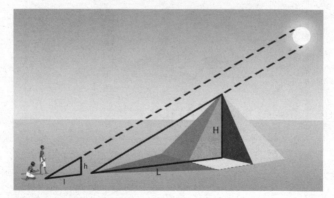

Fig. 19 Similar triangles.

He then measured the shadow ℓ cast by a vertical pole of height h.

Assuming the Sun's rays to be parallel, he reasoned that the two triangles in Fig. 19, though of very different size, would have exactly the same *shape*, and that corresponding sides would therefore be in the same proportion.

In particular, then, he reasoned that

$$\frac{h}{H} = \frac{\ell}{L},$$

and so, having measured the other three lengths, he was able to determine the Great Pyramid's height H.

Fig. 20 Thales, on a Greek postage stamp of 1994.

Today we use the term *similar* to describe triangles which have exactly the same shape, and, as we will see later, they play a major part in some of the most striking theorems of geometry.

Measuring the Earth

According to its Greek roots, the word 'geometry' means, quite literally, 'Earth measurement'. So it seems appropriate to look next at a famous attempt to measure the circumference of the Earth, by Eratosthenes of Alexandria, in about 240 BC.

And, as it happens, he too used the Sun's rays, but in a rather different way.

Eratosthenes knew that, at noon on the longest day of the year, the Sun was directly overhead at his birthplace Syene (modern-day Aswan), because it illuminated the bottom of a deep well there.

He also knew that, at the same time, the Sun made an angle of 7.2° with the vertical at Alexandria, which he took to be 5000 *stades* due north of Syene.

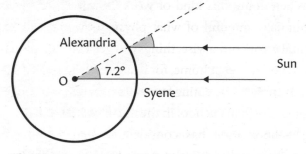

Fig. 21 Measuring the Earth.

Eratosthenes assumed that the Sun was so far from the Earth that its rays arrived parallel. The two shaded angles in

Fig. 21 are then corresponding angles, so the angle at O, the centre of the Earth, must also be 7.2°.

Now, 7.2° is one-fiftieth of 360°, so he reasoned that the circumference of the Earth must be 50 times the distance between Alexandria and Syene, i.e. 250,000 stades.

In truth, this is probably an oversimplification of what Eratosthenes actually did. Moreover, what a *stade* was, as a unit of distance, is also lost; estimates by subsequent scholars put it between 0.15 and 0.2 km, leading to a result for the Earth's circumference somewhere between 37,500 and 50,000 km. (The actual value is about 40,000 km.)

'Practical work', 1929

There is, of course, another, quite different aspect to practical geometry, namely the actual construction of geometrical figures using ruler, compasses, and other tools of the trade.

When doing this kind of work, however, we have to be continually mindful of what physicists would call 'experimental error'; otherwise, things can get a bit ridiculous.

On my shelves at home, for instance, there is an old geometry exercise book, dating from 1929, that once belonged to a pupil at a primary school in the north of England.

The book itself has considerable charm, and consists mostly of simplified Euclid, carried out neatly and well. On the very last page, however, and without any warning at all, we suddenly meet something called 'Practical Work', involving—apparently—some actual *measurement* (Fig. 22).

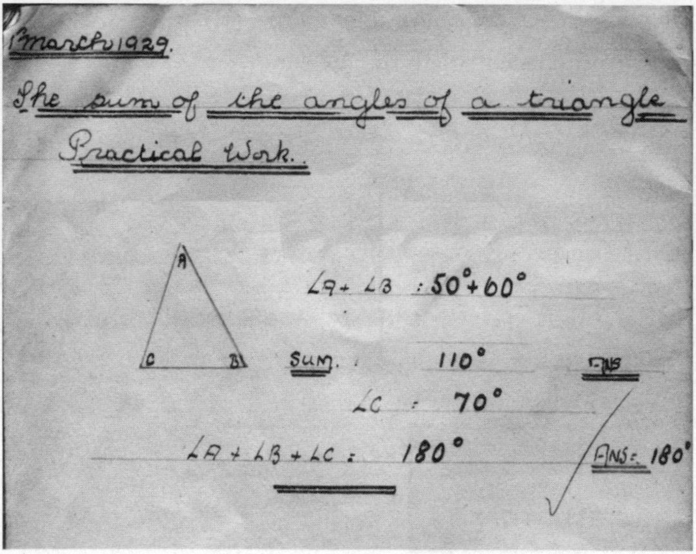

Fig. 22 From a 1929 school exercise book.

Yet, despite the tick of approval from the teacher, there is something faintly absurd about this particular piece of work.

So far as I can determine, the angle A is closer to 45° than 50°, and it looks to me as if the various numbers have simply been cooked up, quite unashamedly, so that the angle-sum comes out 'right'.

Area

Perhaps the oldest geometrical idea of real practical importance is that of *area*, driven largely by problems concerning land.

We begin with a *square* of side 1 unit, and it quickly becomes evident how to calculate the area of a rectangle with sides which are whole numbers (Fig. 23).

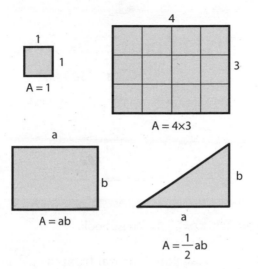

Fig. 23 Area.

This leads us to define the area of a rectangle, more generally, as

$$A = ab,$$

where the side lengths a and b may now be fractional or even irrational.

Introducing a diagonal then bisects the rectangle itself, giving $\frac{1}{2}ab$ as the area of a right-angled triangle.

And, improbable as it may seem, these elementary ideas of area are enough to let us take a first look at one of the most famous—and far-reaching—theorems of all...

Pythagoras' Theorem

There is a surprisingly simple relationship between the lengths of the sides of *any* right-angled triangle (Fig. 24).

And, like so much that is best in mathematics, it is this *generality* that gives the theorem its power.

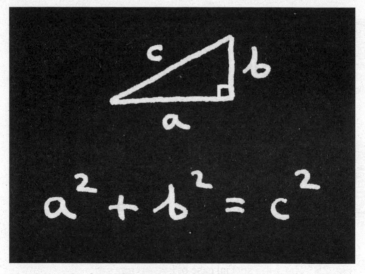

Fig. 24 Pythagoras' theorem.

In Fig. 24, *c* denotes the length of the *hypotenuse*—meaning the side opposite the right angle—while *a* and *b* denote the lengths of the other two sides.

A special case

If the two shorter sides happen to be of length 3 and 4, then $a^2 + b^2 = 9 + 16 = 25$, so c must be 5.

And the idea of a right-angled triangle with sides of length 3, 4, and 5 was known to Babylonian mathematicians (in what is now Iraq) over a thousand years before Pythagoras. One ancient clay tablet, for instance, has a geometrical problem on it equivalent to the following:

> A ladder of length 5 units stands upright, flat against a wall. The upper end slips down a distance 1 unit. How far does the lower end slide out?

Fig. 25 The 3-4-5 special case of Pythagoras's theorem, from John Babington's *Treatise of Geometrie* (1635).

The 3-4-5 right-angled triangle is so well known, in fact, that it is sometimes confused with Pythagoras' theorem itself. Yet, as I have emphasized, it is only one very special case (Fig. 25).

And, in many ways, it isn't the most important special case at all…

Unexpectedly irrational

If the two shorter sides of a right-angled triangle are *equal*, then, according to Pythagoras' theorem, the three sides are in proportion $1:1:\sqrt{2}$ (Fig. 26a).

This was, again, known long before Pythagoras, and a famous Babylonian clay tablet, known as YBC 7289, dating from roughly 1700 BC, shows a square with two diagonals, and various numbers in the mathematical notation of the time (Fig. 26b).

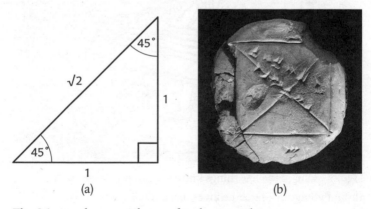

Fig. 26 Another special case of Pythagoras' theorem.

And one of those numbers, representing the ratio of diagonal to side, is, in modern decimal notation:

$$1 \cdot 4142128,$$

which is *within 1 part in a million* of $\sqrt{2}$.

The importance of this particular example stems, in part, from the Pythagorean discovery that $\sqrt{2}$ is *irrational*, so that the ratio of diagonal to side for a square cannot be written exactly as the ratio of two whole numbers.

Fig. 27 One of the few things known with some certainty about Pythagoras is that he investigated the connection between mathematics and music. This woodcut is from Franchino Gaffurio's *Theorica Musicae* (1492).

To put it another way, it is impossible to find a unit of length, *however small*, such that the side and diagonal of a square are both whole numbers.

This was a monumental shock for the Pythagorean world view, and even had a profound effect, centuries later, on the structure of Euclid's *Elements*.

Three proofs of Pythagoras

So far, we have presented Pythagoras' theorem as an unexpectedly simple relationship between three *lengths*.

Evidently, however, $a^2 + b^2 = c^2$ also represents an extraordinary relationship between the *areas* of the three squares that one could, if one wished, construct on the sides of the triangle (Fig. 28).

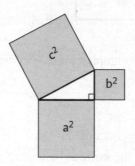

Fig. 28 Another view of Pythagoras' theorem.

And this way of looking at things forms the basis, in fact, of many of the proofs.

A 'proof by picture'

Arrange four copies of the original right-angled triangle as in Fig. 29a. This produces a square with area c^2 in the middle.

Now think of the triangles as white tiles on a dark floor, and move three of them so that they occupy the new positions indicated in Fig. 29b.

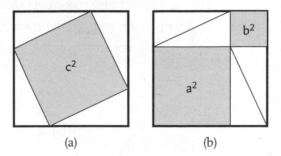

(a)   (b)

Fig. 29 The simplest proof of Pythagoras' theorem?

The floor area *not* occupied by triangles is now $a^2 + b^2$, yet must evidently be the same as before.

So $a^2 + b^2 = c^2$.

'Plain and Easie'

This proof starts with the same diagram, but uses a little algebra instead.

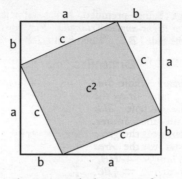

Fig. 30 An algebraic proof.

The area of the large square in Fig. 30 is

$$(a+b)^2 = a^2 + 2ab + b^2.$$

But the large square is made up of the small square (area c^2) and four triangles of area $\frac{1}{2}ab$ each. So the area of the large square is also $c^2 + 2ab$, and therefore

$$a^2 + b^2 = c^2.$$

The earliest clear exposition I have found of this particular proof is in John Ward's 'plain and easie' *Young Mathematician's Guide* of 1707, which was one of the most popular and best-selling mathematics books of its time (Fig. 31).

These things being premised, let us suppose, the *Triangle* *B C H* to be a *Right-angled Triangle*. *Viz.* the Side *C* perpendicular to the Side *B*. Then will $BB + CC = HH$.

Demonstration.

Make a *Square* whose *Side* is $= B + C$, and draw the included *Square* whose *Side* is $= H$, as in the *Scheme*. Then will the *Area* of the great *Square*, be equal to the *Area* of the Four *Triangles* $+ HH$, but the *Area* of each $\triangle = \frac{1}{2}BC$. Per *Lemma* 3. Therefore the $4 \triangle$'s $= \frac{1}{2}BC \times 4 = 2BC$. Consequently, the *Area* of the great *Square* is $HH + 2BC$. *Involve* $B + C$, and it will be $BB + 2BC + CC =$ the *Area* of the great *Square*. Per *Lemma* 2.

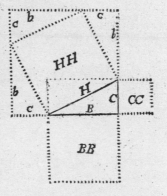

Consequently $HH + 2BC = BB + 2BC + CC$. Per *Axiom* 5. Substract 2BC from both Sides of the *Æquation*, and there will Remain $HH = BB + CC$.   Q. E. D.

Fig. 31 From the *Young Mathematician's Guide* of 1707.

Pythagoras in China?

A similar—but nonetheless different—argument makes use of Fig. 32.

Again there are four copies of the original triangle, each of area $\frac{1}{2}ab$, but they are placed *inside* a square of area c^2 so as to leave a small square of area $(a-b)^2$ in the middle.

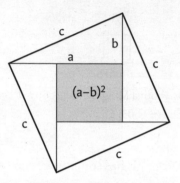

Fig. 32 Another algebraic proof.

So $(a-b)^2 + 2ab = c^2$, and as $(a-b)^2 = a^2 - 2ab + b^2$ this reduces, yet again, to

$$a^2 + b^2 = c^2.$$

And the figure used in this proof can, in fact, be clearly seen in one of the oldest and most famous Chinese mathematical texts, the *Zhou bi suan jing* (Fig. 33).

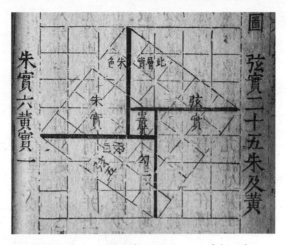

Fig. 33 From a Ming dynasty copy of the *Zhou bi*, printed in 1603.

In consequence, this has sometimes been advanced as one of the earliest 'proofs' of Pythagoras' Theorem. As I understand it, however, the accompanying text is extremely difficult to translate and interpret, and expert scholars on the history of Chinese mathematics are still not agreed on what it really means (see Notes, p. 241).

In any event, as it happens, the proof in Euclid's *Elements* is quite different from anything we have seen so far.

And, some would say, a bit scary…

'In Love with Geometry'?

In John Aubrey's *Brief Lives* there is a famous passage concerning the seventeenth-century philosopher Thomas Hobbes:

> He was 40 yeares old before he looked on Geometry; which happened accidentally. Being in a Gentleman's Library, Euclid's Elements lay open, and 'twas the <u>47 *El. Libri I*</u>. He read the Proposition. By G—, sayd he (he would now and then sweare an emphaticall Oath by way of emphasis) *this is impossible!*

"Show me that again," blurted Janet excitedly.

Fig. 34 Euclid's Proof of Pythagoras' theorem.

Now, *Elements* I.47 is none other than Euclid's statement and proof of Pythagoras' theorem (Fig. 34), and many people find Euclid's diagram rather forbidding. (It certainly terrified me, aged 10, in 1956.)

But Hobbes reacted rather differently, and though he found the theorem itself almost unbelievable, he persisted with the proof, step by step, until at last he was convinced.

And, according to Aubrey, this made him

>...in love with Geometry.

In any event, Euclid's proof is certainly well worth pursuing, and if we are to do this we need, first, to take a closer look at the whole idea of *area*.

The area of a triangle

The general formula for the area of a triangle is shown in Fig. 35.

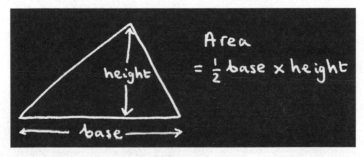

Fig. 35 The area of a triangle.

And we can prove it quite easily by letting the height divide the original triangle into two right–angled ones. As we saw in

Chapter 5, each one has area equal to half the height times *its* base, so by adding the two the result follows.

And if the original arrangement happens to look like Fig. 36b, with an *obtuse*-angled triangle, the result *still* follows;

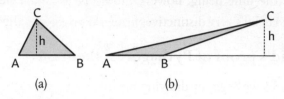

(a)   (b)

Fig. 36 Same area!

we just end up subtracting the areas of two right-angled triangles, instead of adding them.

As a matter of fact, because the two triangles in Fig. 36 have the same height *h* and the same base AB they will have the *same* area, and this will remain true *no matter how far we move* C *to the right*, provided only that we move it always parallel to AB, so that *h* stays the same.

And this was apparently one of the very first results in geometry that impressed the young Isaac Newton, while still an undergraduate at Cambridge.

For, according to a contemporary, he started reading Euclid's *Elements* in 1663, but found the early propositions so obvious that

> he wondered how anybody would amuse themselves to write any demonstrations of them.

Yet, like most of us, perhaps, he didn't find the result in Fig. 36 intuitively obvious at all.

And, interestingly, this particular result was eventually to play a major part in his researches on planetary motion, as we will see later in the book.

For the time being, however, it is just what we need to tackle Euclid's very distinctive proof of Pythagoras' theorem.

Euclid's proof of Pythagoras' theorem

In Fig. 37, we begin by drawing the three squares on the sides of the right-angled triangle, and then draw the line CDE perpendicular to the hypotenuse AB.

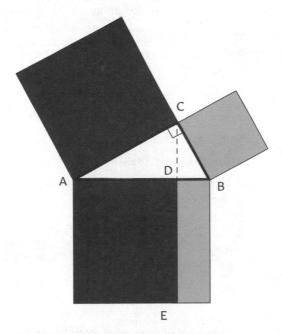

Fig. 37 The idea behind Euclid's proof.

The basic idea is to prove that the areas of the light grey square and light grey rectangle are equal.

By the same argument, the same will be true of the dark grey square/rectangle, too.

In this way, then, the sum of the areas of the two small squares will be equal to the area of the large square on the hypotenuse, and the theorem will be proved.

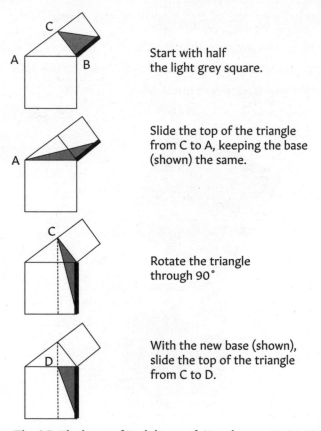

Start with half the light grey square.

Slide the top of the triangle from C to A, keeping the base (shown) the same.

Rotate the triangle through 90°

With the new base (shown), slide the top of the triangle from C to D.

Fig. 38 The heart of Euclid's proof. (See also pp. 12, 36, 179.)

The key, therefore, is to prove the claim about the light grey square and rectangle.

And to do this, we focus essentially on *half* the light grey square (Fig. 38).

The beauty of this argument is that each of the three steps preserves the light grey area, so that the area of the light grey triangle at the end is equal to the area of the light grey triangle at the start. Doubling both then gives the result.

And the only slightly informal step—rotating the triangle through 90° and claiming that it ends up in the position indicated—can be made more formal, if desired, by observing that the two triangles in question are congruent by SAS, the angle between the sides in each case being $90° + \angle ABC$.

371 PROOFS OF **PYTHAGORAS**

In 1927, Elisha Scott Loomis, of Cleveland, Ohio, published a book containing 230 proofs of Pythagoras' theorem . . .

The Pythagorean Proposition

Its Proofs Analyzed and Classified
And
Bibliography of Sources
For Data of
The Four Kinds of Proofs

By
Elisha S. Loomis, Ph. D., LL. B.
Professor Emeritus of Mathematics
Baldwin-Wallace College
Professor Mathematics, Baldwin University, 1885-1895, Head
of Mathematics Department, West High School,
Cleveland, Ohio, 1895-1923,
and
Author of "Original Investigation or How to
Attack an Exercise in Geometry."

ELISHA SCOTT LOOMIS

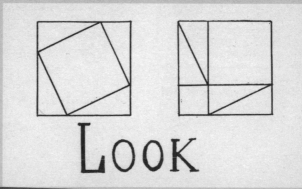

LOOK

By the **2nd** edition (1940) the number had risen to 371, including one by a 16-year old schoolgirl from Indiana ...

From the Indianapolis Star, **28** Oct **1938**

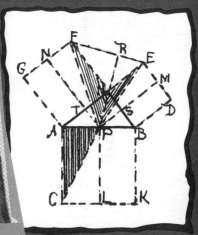

Ann Condit, at Wellesley College in **1944**

(see Notes, p. 241)

8

'Imagine my Exultation, Watson...'

Sherlock Holmes is forever pursuing some master criminal, but not often, I think, with the aid of geometry.

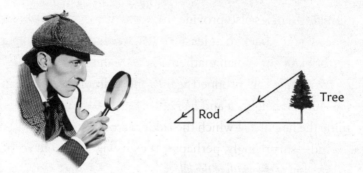

Fig. 39 Sherlock Holmes and similar triangles.

Yet, at a crucial point in 'The Adventure of the Musgrave Ritual', he needs to find the length of the shadow that would have been cast, at a particular time of day, by a long-vanished tree of height 64 feet.

So, having waited until the Sun is at the right position in the sky, he takes a 6-foot fishing rod, measures the length of *its* shadow, and scales up the answer by a factor of 64/6.

And when he examines the actual spot where the shadow of the tree would have ended, he gets rather excited:

> 'You can imagine my exultation, Watson, when...I saw a conical depression in the ground. I knew that it was the mark made by Brunton in his measurements, and that I was still upon his trail.'

Yet this is, of course, just Thales and the Great Pyramid, all over again.

It is just *similar triangles*.

A problem with ladders

Similar triangles also provide the key to a very old problem indeed, going back to at least AD 850, when it appeared in a textbook by the Indian mathematician Mahavira.

Two ladders are propped against the walls of an alley, as in Fig. 40. The heights *a* and *b* are given, and we have to determine the height *h* at which the ladders cross.

And—surprisingly, perhaps—the answer turns out to be *independent of the width of the alleyway*.

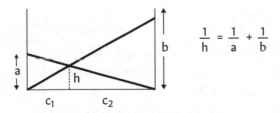

$$\frac{1}{h} = \frac{1}{a} + \frac{1}{b}$$

Fig. 40 A problem with ladders.

Yet to show this, all we have to do is introduce the distances c_1 and c_2, as shown, so that $c = c_1 + c_2$ is the width of the alleyway.

Then, by similar triangles,

$$\frac{h}{b} = \frac{c_1}{c}$$

and

$$\frac{h}{a} = \frac{c_2}{c}.$$

Adding, and dividing by h, then gives the result in Fig. 40.

Pythagoras by similar triangles

Rather more significantly, perhaps, similar triangles provide some of the most elegant proofs of Pythagoras' theorem.

In fact, Albert Einstein once recalled inventing such a proof 'after much effort' as a young boy. Sadly, however, we cannot be sure of the exact details, because there are two major ways in which such a proof can go.

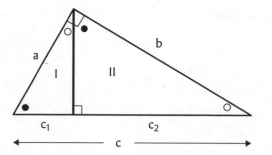

Fig. 41 Pythagoras' theorem by similar triangles.

They both start by dropping a perpendicular from the right angle onto the longest side, or hypotenuse (Fig. 41). Somewhat remarkably, all three triangles are then similar, because they all contain the same three angles.

A neat proof

We observe, next, that *hypotenuses* of these similar triangles will be in the same proportion as the smaller sides.

From Triangle I and the original, large, triangle, we therefore have

$$\frac{a}{c} = \frac{c_1}{a}, \ \text{so} \ a^2 = cc_1.$$

And, in the same way, from triangle II and the large triangle:

$$\frac{b}{c} = \frac{c_2}{b}, \ \text{so} \ b^2 = cc_2.$$

Adding, and recalling that $c_1 + c_2 = c$, then gives the result.

An even neater one?

Areas of similar triangles are in proportion to the *squares* of their corresponding sides.

If we focus, then, entirely on *hypotenuses* in Fig. 41, the area of triangle I must be a fraction a^2/c^2 of the area of the large triangle.

The same argument applied to triangle II produces a fraction b^2/c^2.

The two fractions must evidently add up to 1, so $a^2 + b^2 = c^2$.

Similarity and area

So far in this book, we have treated the ideas of similarity and area quite separately, but they are in fact related.

To see this, draw a rectangle ABCD, and a diagonal AC. Then pick a point E on the diagonal and draw lines through it parallel to the sides (Fig. 42).

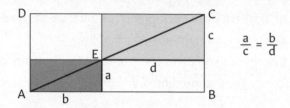

Fig. 42 Area and similarity.

The two right-angled (bold) triangles are similar, because they share the same three angles. So, according to Thales, their sides will be in the same proportion.

And however obvious this may seem, it is possible to deduce it, if we wish, using area.

Note, first, that the two shaded rectangles are bisected by the diagonal AEC, so that the two light grey triangles have equal area, and the same is true of the two dark grey triangles.

But the diagonal AEC also bisects the *large* rectangle, so the triangles ACD and ACB have equal area.

It then follows at once, by subtraction, that *the two (unshaded) rectangles left over must have equal area.* So $ad = bc$, and dividing by cd gives the result:

$$\frac{a}{c} = \frac{b}{d}.$$

The bigger picture

Ideas of similarity are very powerful, but we have applied them, so far, only to right-angled triangles.

And our work so far on congruence has a serious draw-back, too: we have, at present, only one way (SAS) of identifying it.

It is time, then, to take a closer, and more general, look at congruence and similarity as a whole.

9

Congruence and Similarity

The 'Dam Buster' raids of 1943 have gone down in history as one of the most extraordinary episodes of World War II.

But in order for the 'bouncing bombs' to skip, repeatedly, over the water, and reach their target, they had to be dropped from an aircraft flying at a precise height of just 60 feet.

To achieve that, the ground crew mounted one angled spotlight in the aeroplane's nose and another nearer its tail, so defining a unique triangle with the required height (Fig. 43).

Fig. 43 The dam busters.

And when the two spots of light on the water's surface coincided, the bomb was released.

So SAS is not the only way of specifying the exact shape and size of a triangle, and the method we have just seen is called—not surprisingly—ASA.

There is in fact a third major method, called SSS, in which we simply specify the lengths of all three sides. This is, perhaps, the least 'obvious' of the three, but imagine, if you will, taking a line of given length and drawing a circle of some given radius about each end (Fig. 44).

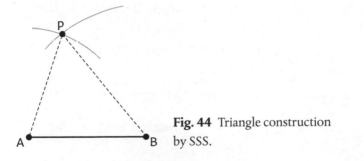

Fig. 44 Triangle construction by SSS.

It is evident, I think, that if the two circles meet *at all*, they will meet, on any particular side of the original line, at a *unique* point P.

We now have *three* ways, then, of defining the precise shape and size of a triangle, and these lead directly to three ways of identifying, and exploiting, congruence.

Congruence

The three major tests for congruent triangles are shown in Fig. 45.

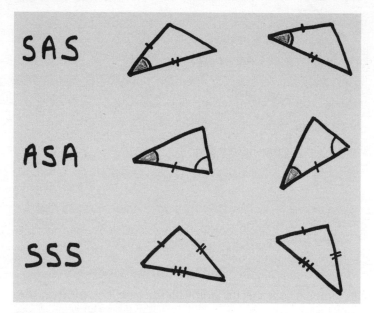

Fig. 45 Tests for congruence.

This is, perhaps, the place to stress that the angle in 'SAS' must be *between* the two sides, just as the side in 'ASA' must be 'between' the two angles.

But, having made this point, I would like to move swiftly on to two examples of the congruence tests in action.

The reflection of light

When light is reflected at a plane mirror, experiment shows that the incident and reflected rays make equal angles with the mirror.

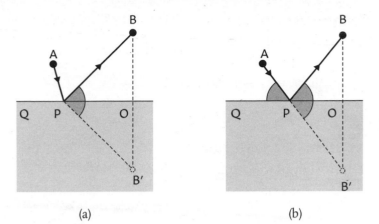

(a)   (b)

Fig. 46 Finding the shortest path.

And in about AD 100, Heron of Alexandria realized that, in doing this, light takes—somewhat mysteriously—the *shortest possible path*.

To see this, imagine, first, the light travelling from a specific point A to a specific point B via a *general* point P on the mirror (Fig. 46a)

Now draw BO perpendicular to the mirror, and extend it to B′, where OB′= OB. The triangles POB and POB′ will then be congruent by SAS (the angle in question being 90°).

So, wherever P is on the mirror,

$$PB = PB' \quad \text{and} \quad \angle BPO = \angle B'PO$$

Suddenly, then, the problem of picking P to minimize AP + PB is the same as picking P to minimize AP + PB′—and we do that, surely, by making APB′ a straight line (Fig. 46b).

And in that particular case, $\angle APQ = \angle B'PO$, because they are opposite angles. As $\angle B'PO = \angle BPO$ *anyway*, it follows that, in the shortest-path case, the incident and reflected rays make equal angles with the mirror.

Parallelograms

A parallelogram is defined simply as a four-sided plane figure in which opposite sides are parallel.

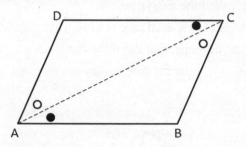

Fig. 47 A parallelogram.

But opposite sides are also *equal*, and we can prove this using ASA.

The key step is to draw in a diagonal such as AC in Fig. 47. The angles • are equal (alternate angles), and the angles o are also equal, for the same reason.

The triangles ABC and CDA are then congruent by ASA.

So, in particular, AB = CD and BC = DA, which is what we were trying to show.

Describing congruence and similarity

Note, incidentally, that in referring to triangles ABC and CDA just now, I was trying to communicate, *through the order of the lettering*, not only the congruence but also *which points corres-pond to which*, and I will try to follow this practice with both congruence and similarity throughout the book.

Similarity

Similar triangles have *exactly the same shape*, and therefore the same three angles.

And, in consequence, their sides are all in the same pro-portion k, called the *scale factor*.

Congruence is a special case of similarity, with $k = 1$, so it is no surprise, perhaps, that the three main tests for similarity (Fig. 48) mirror those for congruence very closely.

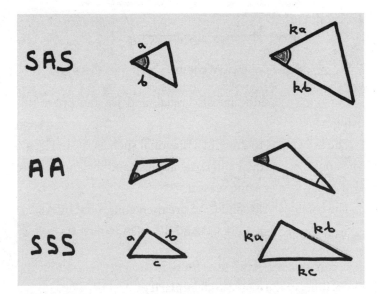

Fig. 48 Tests for similarity.

And, as with congruence, I would like to offer, straight away, two examples of the tests in action.

The mid-point theorem

Take any triangle and join the mid-points of two of its sides.

The resulting line is then *parallel to the third side* (Fig. 49).

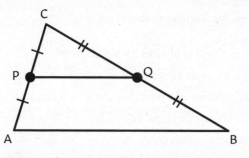

Fig. 49 The mid-point theorem.

To see this, we compare the two triangles PCQ and ACB. They have the angle C in common, and $PC = \frac{1}{2}AC$ while $QC = \frac{1}{2}BC$.

They are therefore similar by SAS, with scale factor $k = \frac{1}{2}$.

So all their angles must correspond, and therefore, in particular, $\angle CPQ = \angle CAB$.

So PQ and AB are parallel.

Varignon's theorem

Take any quadrilateral—i.e. four-sided figure—and join the mid-points of the sides by straight lines, in order (Fig. 50).

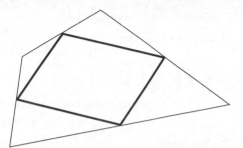

Fig. 50 Varignon's theorem.

Then the result is always a *parallelogram*!

And yet, as soon as we draw in one of the diagonals of the original figure, we see why, because two applications of the mid-point theorem show that GH and FE are both parallel to DB, and must therefore be *parallel to each other* (Fig. 51). And an equivalent argument, using the other diagonal, shows that EH is parallel to FG.

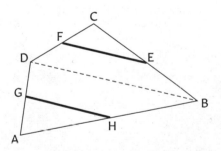

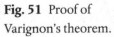

Fig. 51 Proof of Varignon's theorem.

Rarely in mathematics, I think, does such an apparently mysterious result become so 'obvious', so quickly, when you look at it the right way.

THE GOLDEN RATIO

$$\Phi = \frac{1 + \sqrt{5}}{2}$$

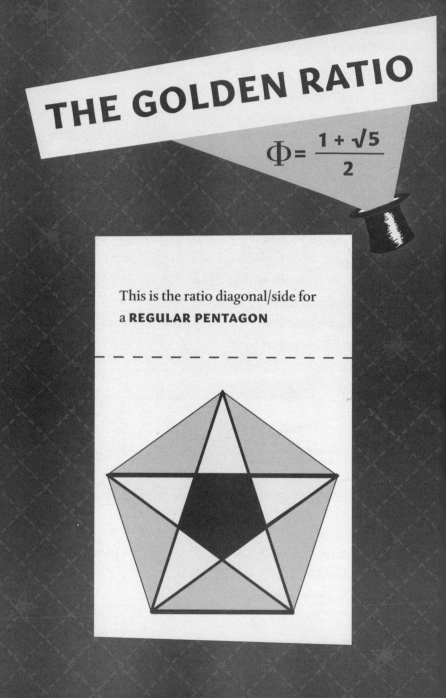

This is the ratio diagonal/side for a **REGULAR PENTAGON**

PROOF Each diagonal is parallel to one of the sides, so ABDC is a **parallelogram**. So DC = 1 and DE = Φ −1

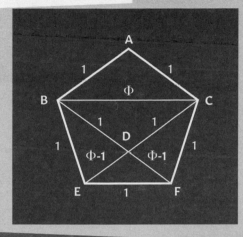

The triangles ABC and DEF are **similar**, so BC/EF = AB/DE, which leads to

$$\Phi^2 = \Phi + 1$$

The positive root of this quadratic is

$$\frac{1 + \sqrt{5}}{2}$$

Conversely...

In ancient Egypt, there were surveyors of land known as 'rope-stretchers', and it is sometimes claimed that they used the 3–4–5 special case of Pythagoras' theorem to create right angles (Fig. 52).

Well, they *didn't*.

Fig. 52 Ancient Egypt?

This is because Pythagoras' theorem says that *if* a triangle has a right angle then there is a certain relationship between the lengths of the three sides.

But what the rope-stretchers would have needed—if they did anything like this at all—is the exact opposite.

In short, they would have needed not Pythagoras' theorem, but its *converse*.

And, more generally in mathematics, it is always import-ant to distinguish carefully between any statement

P *implies* Q

and its converse

Q *implies* P,

which, like the original statement itself, may or may not be true.

The converse of Pythagoras' theorem

So, is the converse of Pythagoras' theorem true? In particular, is it true that in a 3–4–5 triangle the angle opposite the long-est side must necessarily be 90°?

Well, actually it is, and the proof is very simple—almost laughably so. All we have to do is draw a *second* triangle with sides of length 3 and 4, and an angle of 90° between them (Fig. 53b).

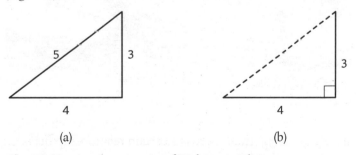

(a)   (b)

Fig. 53 Proving the converse of Pythagoras' theorem.

By Pythagoras' theorem itself, the length of the dashed line in Fig. 53b will be 5. The two triangles are then congruent by SSS, so the angle opposite the longest side in Fig. 53a must, indeed, be 90°.

Proof by contradiction

Proving the converse of some theorem is not usually so straightforward, and one very helpful device is the idea of proof by contradiction, or *reductio ad absurdum*.

The idea is to assume, at the very outset, that what you are trying to prove is *false*, and then show that this would lead to some contradiction or absurdity.

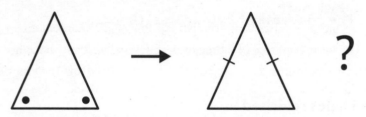

Fig. 54 Converse of the isosceles triangles theorem.

Isosceles triangles, for instance, provide a good example. We saw earlier that if a triangle is isosceles then its base angles are equal. But what about the converse (Fig. 54)?

It is in fact true, and Euclid proves it in the *Elements* by contradiction.

He starts, then, with ∠ABC = ∠ACB in Fig. 55, but assumes that AB and AC are *not* equal. One of them must therefore be larger than the other, and he lets AB denote the larger. Then he introduces the point D such that DB = AC, and draws in the line CD.

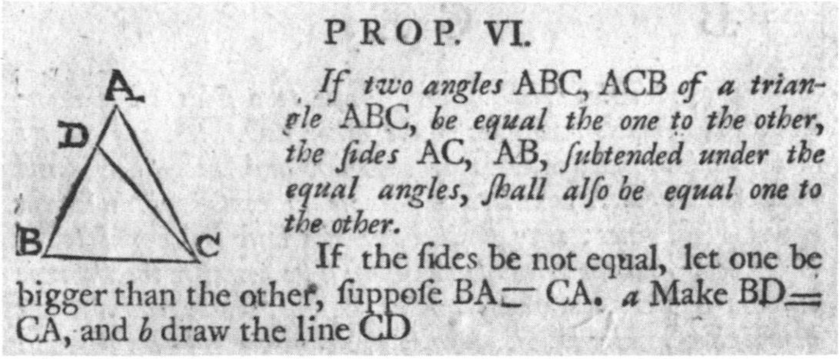

PROP. VI.

If two angles ABC, ACB of a triangle ABC, be equal the one to the other, the sides AC, AB, subtended under the equal angles, shall also be equal one to the other.
    If the sides be not equal, let one be bigger than the other, suppose BA ⌐ CA. *a* Make BD = CA, and *b* draw the line CD

Fig. 55 A proof by contradiction in Barrow's edition of Euclid's *Elements* (1732).

The triangles DBC and ACB are then congruent by SAS.

But this is absurd, because triangle DBC is only *part* of triangle ACB!

The original assumption that AB and AC are unequal must therefore be wrong. So they *are* equal, which proves the converse theorem.

Thales revisited

A few years ago, I was enthusing about Thales' theorem (Fig. 56) at some conference, and two members of the

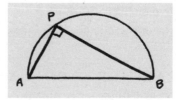

Fig. 56 Thales' theorem.

audience came up to me afterwards and claimed that I was making a bit of a fuss, and that the result was really rather 'obvious'.

After all, they said, it's fairly obvious from symmetry that we can take any rectangle and place it in a circle of suitable size so that all four corners lie on the circle itself (Fig. 57a).

And it's also fairly obvious—again from symmetry—that the diagonals of the rectangle meet at the centre of the circle (Fig. 57b).

So if we now rub out half the diagram we get Fig. 57c, with the dotted line as a diameter.

And that's Thales' theorem, isn't it?

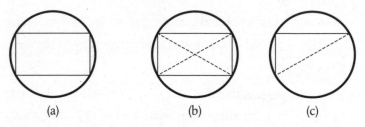

(a)   (b)   (c)

Fig. 57 A proof of ... what, exactly?

Well ... no ... I don't think it is.

Loosely speaking, with Thales' theorem, we start with a diameter and end up with a right angle, not the other way round.

More precisely, let A and B be two different points on a circle, and let P be a general point in the plane, not coincident with either A or B. And consider, if you will, the following three statements:

(a) P is on the circle

(b) AB is a diameter

(c) ∠APB is a right angle.

Thales' theorem says that (a) + (b) implies (c).

And, put like this, *two* possible converses come to mind.

One is that (a) + (c) implies (b), and it seems to me that the foregoing argument is an informal proof of this.

The other possible converse, and the one I will refer to in future as *the* converse, because of its practical value, is (b) + (c) implies (a).

We will now prove this by a powerful device, namely coupling proof by contradiction *with the original theorem itself.*

The converse of Thales' theorem

Draw a circle with AB as diameter, and let P be some point not on AB.

We want to show that if ∠APB = 90°, then P lies on the circle.

Begin, then, by supposing that ∠APB = 90°, but that P *doesn't* lie on the circle. There are then two cases to consider.

Suppose, first, that P lies inside the circle (Fig. 58).

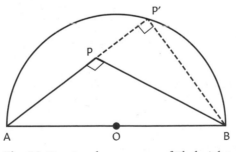

Fig. 58 Proving the converse of Thales' theorem.

We then extend the line AP so that it meets the circle at some point P′, say, and we join P′B.

Now, ∠APB = 90° by assumption, and ∠AP′B = 90°, *by Thales' theorem itself.*

So PB and P′B are both perpendicular to AP′, and must therefore be parallel.

But they meet, of course, at B, giving a contradiction.

So the assumption that P is inside the circle must be wrong.

The same argument, with a slightly different diagram, shows that P can't be outside the circle either.

So P must lie on the circle itself.

An alternative approach

The method we have just introduced is a powerful and *general* one, but, as it happens, there is on this occasion a simpler alternative that avoids 'contradiction' altogether.

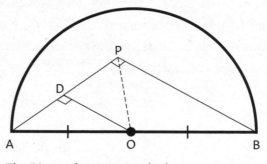

Fig. 59 An alternative method.

In Fig. 59, with ∠APB = 90°, draw OD parallel to BP.

Triangles AOD and ABP are then similar (by AA), with a scale factor of 2 (because AB = 2 × AO), so D is the mid-point of AP.

Draw OP. Triangles ODA and ODP are then congruent by SAS, so OP = OA, and P therefore lies on the circle.

(Yet another proof of the converse of Thales' theorem can be found in Notes, p. 243.)

And, just in case you should wonder, I make no apology for returning, from time to time in this book, to the very first theorem in geometry that made a real impression on me, aged 10.

That's because, for all its elegance, and element of surprise, Thales' theorem is just one special case of something even more general and far-reaching . . .

Circle Theorems

The main circle theorem, from which so much else follows, can be loosely stated as:

The angle at the centre is twice the angle at the circumference,

where the angles in question stand on the same arc of the circle (Fig. 60a).

And it has an immediate and quite extraordinary consequence: all the angles like $\angle APB$ and $\angle AP'B$ in Fig. 60b are *equal*, because *they are all half the angle at the centre!*

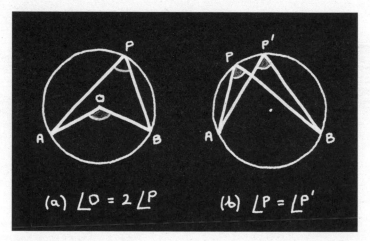

Fig. 60 Circle theorems.

Why, then, is the remarkable theorem in Fig. 60a true?

The main theorem

In Fig. 61, triangle AOP is isosceles, because OA = OP (radii). The two 'base angles' a are therefore equal. Moreover, $\angle AOP = 180° - 2a$, so if we extend the straight line PO to Q, then $\angle AOQ = 2a$.

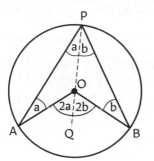

Fig. 61 Proof of the main theorem.

In a similar way, the two angles b are equal, and $\angle BOQ = 2b$. Adding then gives the result

$$\angle AOB = 2\angle APB.$$

In truth, there is in fact another case to consider, in which, for example, P lies so far to the right that AP crosses OB. Happily, however, the argument goes through almost as before; we just end up subtracting two angles instead of adding them.

Fig. 62 The main circle theorem, on a Guinea–Bissau postage stamp.

The theorem in Fig. 60b then follows at once, and this, in turn, leads to two further remarkable theorems…

Four points on a circle

If four points happen to lie on a circle, then opposite angles of the associated quadrilateral add up to 180° (Fig. 63).

One way of seeing why is to note first that the angles s are equal, for they stand on the same arc AB. The angles t are also equal, for a similar reason.

But from triangle ABC we have $s + t + \angle ABC = 180°$. So

$$\angle ADC + \angle ABC = 180°.$$

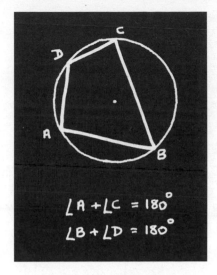

$$\angle A + \angle C = 180°$$
$$\angle B + \angle D = 180°$$

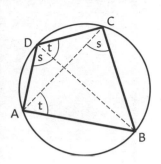

Fig. 63 Four points on a circle.

Arguably, however, the *converse* of this result turns out to be even more important:

If the opposite angles of a quadrilateral add up to 180°, then the four points concerned lie on a circle.

The standard proof of this is by contradiction, and proceeds in exactly the same spirit as that for the converse of Thales' theorem.

Intersecting chords

If two chords AB, CD meet at a point P, then

$$AP. BP = CP. DP$$

To see why, note that in Fig. 64, the angles *s* are equal, and the angles *t* are also equal.

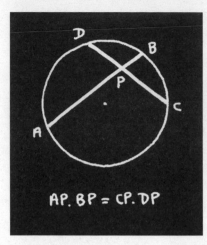
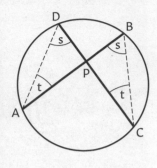

AP. BP = CP. DP

Fig. 64 Intersecting chords.

Triangles APD and CPB are therefore similar, by 'AA', so AP/CP = DP/BP, whence the result. And the same argument works equally well if the two chords happen to meet *outside* the circle (Fig. 65).

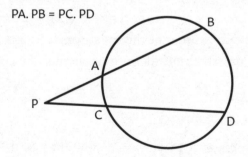

PA. PB = PC. PD

Fig. 65 An external intersection.

Off at a Tangent

'Am I going to die?' said Tangent, his mouth full of cake.

Evelyn Waugh, *Decline and Fall* (1928)

I don't know why it is, exactly, that non-mathematicians often find the idea of a *tangent* somehow amusing.

In Evelyn Waugh's classic novel for instance, Tangent is the hapless and dim-witted son of Lord and Lady Circumference, and gets shot in the foot with a starting pistol on School Sports Day.

Yet there's nothing funny, really, about a tangent.

And all we have to do to get one is to pick some point P on a circle and draw the straight line through P that is at right angles to the radius OP. All the other points on that line will

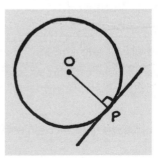

Fig. 66 The tangent to a circle.

then be further from O than P is, and will therefore lie outside the circle.

In other words, this tangent line will just *touch* the circle at P, whence its name.

The secant–tangent theorem

Figure 67, with a tangent PT, shows a classical theorem dating back to Euclid.

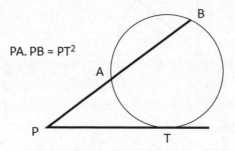

$$PA.PB = PT^2$$

Fig. 67 The secant–tangent theorem.

The simplest, but slightly informal, way of seeing why it is true is, I think, to imagine the 'secant' line PAB (so-called because it cuts the circle) rotating gradually clockwise about P. The product PA.PB will remain constant in the process, by Fig. 65, and as we can eventually make both A and B as close to T as we like, that constant must be PT^2.

(A more formal proof can be found in Notes, p. 244.)

And while this theorem may seem a little obscure, it does in fact have a very practical application...

Measuring the Earth (again!)

The idea here is to deduce the radius of the Earth, R, by climbing a mountain of known height h and estimating the distance D to the horizon.

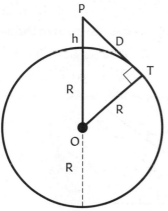

Fig. 68 Measuring the Earth.

If, in Fig. 68, we draw in the tangent line PT, the secant–tangent theorem tells us at once that

$$h(2R+h)=D^2.$$

In practice, h will be so small in comparison with $2R$ that this is sensibly approximated by $2Rh = D^2$, so that

$$R \approx \frac{D^2}{2h}.$$

And this is essentially what the Persian astronomer and mathematician Al-Biruni did, in about 1019, obtaining a value for R of 3939 miles.

This was astonishingly accurate for the time, and within 1% of the currently accepted value.

Looking at Euclid

Another nice example of tangents in action is a curious problem first posed in 1471—in a slightly different way—by the German astronomer Johann Müller, better known as Regiomontanus.

Imagine, if you will, that you are looking up, with due reverence, at an enormous statue of Euclid (Fig. 69).

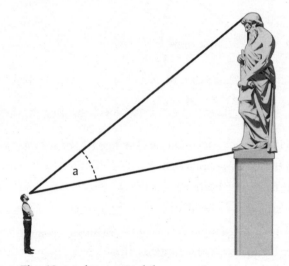

Fig. 69 Looking at Euclid.

Clearly, if you stand too far away, your viewing angle *a* will be very small.

But it will also be small if you stand too close, because you will then be viewing Euclid very obliquely.

So, where should you stand to make the viewing angle *a as large as possible?*

To find the answer, draw a horizontal line at eye level. Then draw the (unique) circle through the top and bottom of Euclid himself that just touches that line (Fig. 70a).

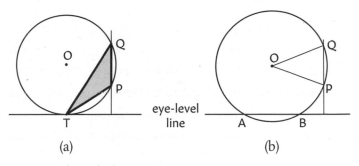

(a)   (b)

Fig. 70 What's the best view?

Viewing from the point of tangency T then gives the largest possible value of the viewing angle *a*.

This is because any other point on the eye-level line, such as A or B in Fig. 70b, will inevitably lie on a *bigger* circle through P and Q. And the viewing angles there will consequently be smaller, because they will both be $\frac{1}{2}\angle POQ$, which will, itself, have become smaller as a result of O moving further from P and Q.

So, if Euclid's head and feet are at heights h_1 and h_2 above your eye level, the secant–tangent theorem gives the optimum viewing distance as $\sqrt{h_1 h_2}$.

A good try?

As it happens, the same problem arises—in principle—in rugby, whenever a player has to select a point on the try-line (Fig. 71), for the viewing angle of the gap between the posts can be maximized in the same way.

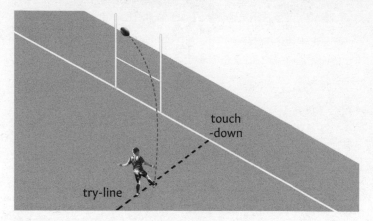

Fig. 71 Another maximization problem?

But there must, I think, be other practical considerations, because I once had the opportunity to present this strategy to a former England international rugby player, and he was deeply unimpressed.

From Tangents to Supersonic Flow

When an object moves faster than the speed of sound, the disturbance to the air is confined to the region behind a v-shaped 'shock wave' (Fig. 72).

Fig. 72 Supersonic flow.

And, somewhat surprisingly, it is possible to infer from a still photograph such as this how fast the object is moving.

More precisely, we can determine the *Mach number*

$$M = \frac{v}{c},$$

where v is the speed of the object, and c the speed of sound.

To do this, however, we need some tangents.

And a bit of trigonometry…

Trigonometry

This branch of the subject begins with the definition of the *sine* and *cosine* of an angle θ, which we do by constructing a right-angled triangle (Fig. 73).

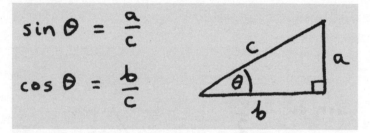

Fig. 73 Definition of $\sin \theta$ and $\cos \theta$.

It is worth noting at once that these definitions are only made possible by the whole idea of similar triangles, because the ratios a/c and b/c depend only on the exact *shape* of the right-angled triangle and not on its size.

The quantities $\sin \theta$ and $\cos \theta$ are not independent; Pythagoras' theorem tells us that $(\sin \theta)^2 + (\cos \theta)^2 = 1$, and a universal shorthand for this is

$$\sin^2\theta + \cos^2\theta = 1.$$

Two values of θ are particularly important, for reasons of symmetry.

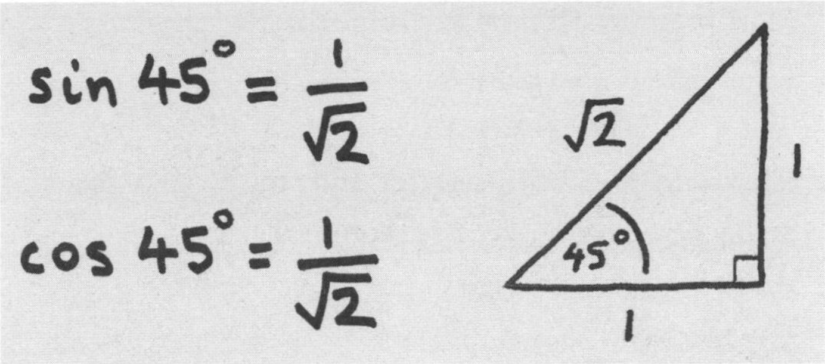

$$\sin 45° = \frac{1}{\sqrt{2}}$$

$$\cos 45° = \frac{1}{\sqrt{2}}$$

Fig. 74 A special case.

The first is 45°, which leads to an *isosceles* right-angled triangle that we have encountered already (Fig. 74).

The other is 60°, in which case we can determine sin θ and cos θ by bisecting an *equilateral* triangle, i.e. one with all three sides equal (Fig. 75).

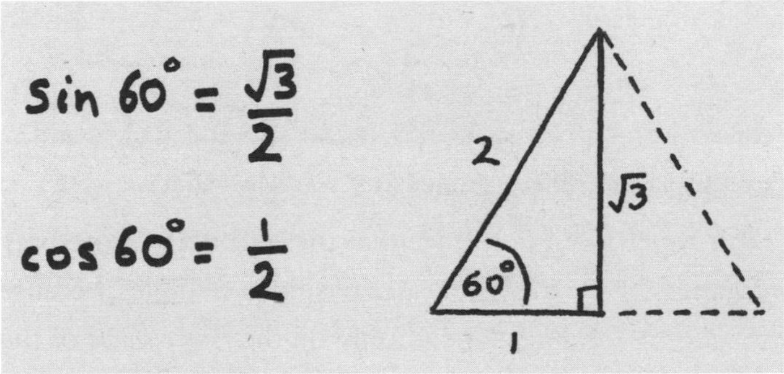

$$\sin 60° = \frac{\sqrt{3}}{2}$$

$$\cos 60° = \frac{1}{2}$$

Fig. 75 Another special case.

Such a triangle is isosceles in three different ways, so *all* its angles are equal, and each one must therefore be 60°. And if we choose a side length of 2, for convenience, the outcome is as shown, with the $\sqrt{3}$ emerging from an application of Pythagoras' theorem.

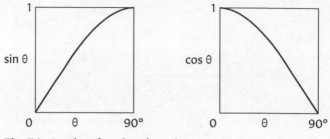

Fig. 76 Graphs of sin θ and cos θ.

Rather more ingenuity (and sheer hard work) is needed to determine sin θ and cos θ for other values of θ, but Fig. 76 shows the overall results in graphical form.

And yet... what has all this got to do with supersonic flow?

Supersonic flow

Recall, first, the idea of *Mach number*

$$M = \frac{v}{c},$$

where v is the speed of the object and c the speed of sound.

Now, as the object passes any particular fixed point P in space, it generates a sound wave which travels outward at speed c, so that at time t later that particular acoustic disturbance is confined to a circle around P of radius ct (Fig. 77).

By that time, the object itself will have travelled a distance vt from P.

So, if $v > c$, the object will now be *outside* the circle in question, and on drawing tangents from it we find that the angle θ is such that sin θ = c/v.

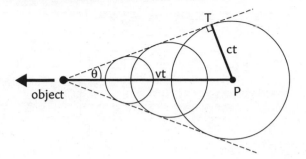

Fig. 77 Tangents in action.

In this way, then, the total acoustic disturbance—due to all such waves—will be confined to a v-shaped wedge of half-angle θ, where

$$\sin\theta = \frac{1}{M},$$

and there will be no disturbance at all outside this region.

So, in the case of Fig. 72, where θ is roughly 45°, we deduce that the object must be moving with a Mach number M of roughly $\sqrt{2}$, or 1·4.

GALILEO AND THALES' THEOREM

In around 1600, **Galileo** conducted his famous experiments on motion down an **inclined plane**.

Galileo's **chord theorem:** motions down all inclined planes such as AB, AC etc. **take the same time:**

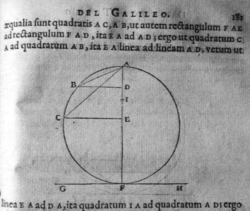

The modern explanation is that the **component** of gravity down the plane is proportional to cos A, so very small if the plane is almost horizontal.

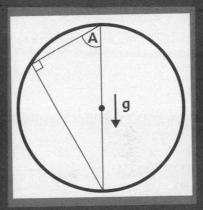

But, by **Thales' theorem**, the **length** of the plane is also proportional to cos A, and the two effects **cancel out**.

What is π, Exactly?

The circumference of any circle is proportional to its diameter, and this allows us to define the special number

$$\pi = \frac{\text{circumference}}{\text{diameter}}.$$

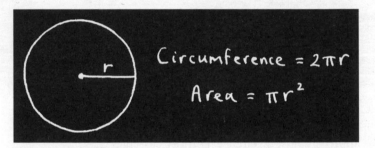

Fig. 78 Circles and π.

Moreover, the diameter is twice the radius r, so the formula for the circumference in Fig. 78 follows directly from our definition of π.

But what about the formula for the area? Why is *that* true?

The area of a circle

We will use an idea due, essentially, to Archimedes, and inscribe within the circle a polygon having N equal sides (Fig. 79).

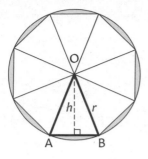

Fig. 79 An inscribed polygon.

This will consist of N equal triangles, such as OAB, where O is the centre of the circle.

Now, the area of each such triangle will be $\frac{1}{2}$ its height h times its base AB. The total area of the polygon will be N times this, i.e. $\frac{1}{2} \times h \times (AB) \times N$.

But $(AB) \times N$ is the length of the perimeter of the polygon, so

$$\text{area of polygon} = \frac{1}{2} \times h \times (\text{perimeter of polygon}).$$

Finally, consider what happens as N gets larger and larger, so that the polygon has shorter and shorter sides, and approximates the circle ever more closely (Fig. 80).

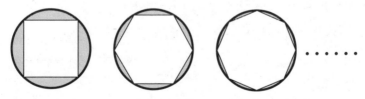

Fig. 80 Closer and closer...

Plainly, h will get ever closer to the radius of the circle, r, and the perimeter of the polygon will get ever closer to the circumference of the circle $2\pi r$.

In short, then, by taking N large enough we can make the area of the polygon as close as we like to $\frac{1}{2} \times r \times 2\pi r$.

And that is why the area of the circle itself is πr^2.

The volume of a sphere

Using broadly similar ideas, Archimedes proved the result in Fig. 81 for the volume of a sphere.

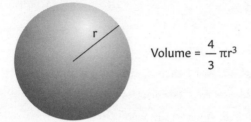

$$\text{Volume} = \frac{4}{3}\pi r^3$$

Fig. 81 The volume of a sphere.

To use any of these formulae in practice, however, we plainly need to know something about the actual *numerical value* of the constant π.

Approximating π

Archimedes started by taking a circle and then calculating the perimeters of inscribed *and circumscribed* hexagons (Fig. 82).

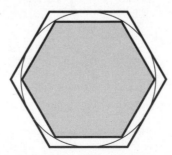

Fig. 82 Approximating π.

The circumference of the circle must be greater than the first of these, but less than the second, and with the help of some ideas from Chapter 13 this leads to

$$3 < \pi < 2\sqrt{3}.$$

But he then doubled the number of sides, repeatedly, until with 96-gons he obtained the tighter bounds:

$$3\frac{10}{71} < \pi < 3\frac{1}{7},$$

and the upper bound, 22/7, was still being used as a 'practical' approximation to π in my early schooldays in the 1950s.

Packing problems

One application of all these ideas is to *packing problems*, which can be even more fiendish in mathematics than in real life.

Circle packing

Suppose, for instance, that we want to pack a lot of circles as tightly as possible on a flat surface.

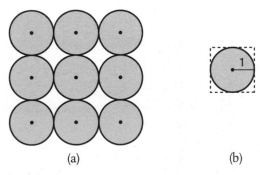

(a)   (b)

Fig. 83 Square packing.

One arrangement that comes to mind, surely, is the one in Fig. 83. If the circles have radius 1, and therefore area π, the plane is completely covered by square units of area $2 \times 2 = 4$, as in Fig. 83b, so

$$\text{square 'packing fraction'} = \frac{\pi}{4},$$

which is about 78%.

But it is fairly obvious, I think, that we can do better with the *hexagonal* arrangement of Fig. 84a, in which each circle is in direct contact with six other circles, rather than just four.

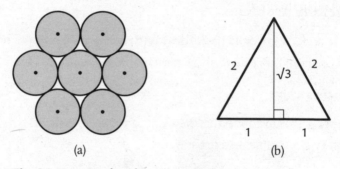

(a)   (b)

Fig. 84 Hexagonal packing.

Each horizontal row is exactly as it was before, but the distance *between* them has decreased. The gap between the lines of centres used to be 2, but is now the height of an equilateral triangle of side 2, which is $\sqrt{3}$ (Fig. 84b).

So the new packing fraction is the square one multiplied by $2/\sqrt{3}$, giving

$$\text{hexagonal packing fraction} = \frac{\pi}{2\sqrt{3}},$$

which is about 90%.

And while we might find it intuitively obvious that the hexagonal packing arrangement is the *best of all*, this was in fact only proved in 1890 by the Norwegian mathematician Axel Thue.

Sphere packing

The corresponding problem with *spheres* is even more difficult.

If we adopt a similar approach to that with circles, we arrive at the packing arrangement of Fig. 85, with each layer arranged hexagonally, and nestling in the troughs of the layer immediately below, as with a stack of oranges at a greengrocer.

And in this case

$$\text{volume packing fraction} = \frac{\pi}{3\sqrt{2}},$$

which is about 74%.

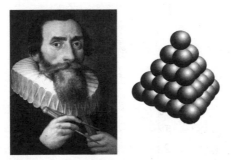

Fig. 85 Johannes Kepler and the packing of spheres.

And in 1611, Johannes Kepler—who was apparently thinking of cannonballs rather than oranges—famously conjectured that no tighter packing was possible.

Yet the first proof of *this* conjecture to survive serious critical scrutiny did not emerge *until 1998!*

The first exact formula for π

We have seen Archimedes' bounds on π, which imply the well-known approximation $\pi \approx 3.14$, but in 1593 Francois Viète obtained the first ever *exact* expression for the numerical value of π, in the form of a product involving an infinite number of terms (Fig. 86).

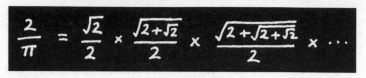

$$\frac{2}{\pi} = \frac{\sqrt{2}}{2} \times \frac{\sqrt{2+\sqrt{2}}}{2} \times \frac{\sqrt{2+\sqrt{2+\sqrt{2}}}}{2} \times \cdots$$

Fig. 86 The first ever exact formula for π.

To get it, Viète started with an inscribed *square* (Fig. 87a), and repeatedly doubled the number of sides—without ever stopping.

To see the effect of this doubling, suppose that at a certain stage in the process one side of the polygon, PQ, makes an angle 2A at the centre (Fig. 87b).

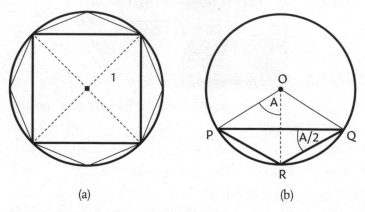

(a)   (b)

Fig. 87 Doubling the number of sides.

Then, at the next stage, PQ is replaced by PR + RQ, and the main circle theorem of Chap. 11 then implies that $\angle PQR = \frac{1}{2}A$.

This means that each time we double the numbers of sides, the current perimeter of the inscribed polygon gets multiplied by a factor

$$\frac{1}{\cos\frac{1}{2}A}.$$

And A itself will, of course, be reduced by a factor of 2 when we apply the next stage of the process.

So, if the circle has radius 1, the starting square will have perimeter $4\sqrt{2}$ and the starting value of A will be 45°. The circumference of the circle itself will be 2π, so by repeatedly doubling the number of sides we get

$$2\pi = 4\sqrt{2} \cdot \frac{1}{\cos\dfrac{45°}{2}} \cdot \frac{1}{\cos\dfrac{45°}{4}} \cdot \frac{1}{\cos\dfrac{45°}{8}} \cdots$$

To finish things off we appeal to a useful little result from trigonometry for the cosine of *half* a given angle:

$$\cos\frac{\theta}{2} = \frac{\sqrt{2+2\cos\theta}}{2}$$

(see Notes, p. 246). As $\cos 45° = 1/\sqrt{2}$, repeated application of this, and a little rearranging, leads directly to Viète's extraordinary result.

The Story of the Ellipse

On 24 October 1811, a mathematics student at Oxford called George Chinnery wrote in his diary

> That odious and abominably tedious book called the Conic Sections.... is forced upon us merely because the author of the treatise is alive and resident in the University.

And I find this rather sad, because a conic section is what we get by taking a plane slice through a circular cone (Fig. 88).

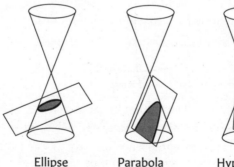

Ellipse Parabola Hyperbola

Fig. 88 Conic sections.

In particular, if we slice through the cone at a fairly shallow angle, the result is a closed curve called an ellipse.

And even by 1811, whether George Chinnery knew it or not, the ellipse had already proved to be one of the most important curves in the history of science.

The ellipse

There is in fact another, completely different way of making an ellipse.

Mark out two fixed points, H and I in Fig. 89, and run a loop of string round them. Then move the point E, while keeping the string taut; this will trace out an ellipse.

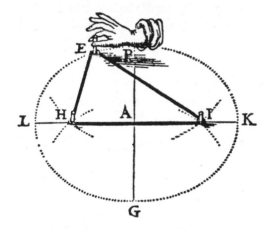

Fig. 89 An ellipse, from van Schooten's *Exercitationum Mathematicarum* (1657).

If the loop of string is long, the ellipse will be almost circular, but if it is short enough, so that it fits rather tightly about the two fixed points, the ellipse will be long and thin.

And the two fixed points are called *focal points*, for good reason...

The reflection property

If we make the perimeter of the ellipse a *mirror,* and place a source of light at one focal point, the light will be reflected by the mirror *to the other focal point.*

This is because, in Fig. 90, the lines PF and PF′ *make equal angles with the tangent.*

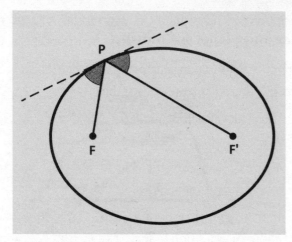

Fig. 90 The reflection property of an ellipse. F and F′ are the focal points.

And to see why this property holds, imagine that we are moving around the ellipse, towards the fixed point P, keeping the string taut.

Now, if we suddenly decide to 'fly off at a tangent' at P, we can only do so if we can somehow *lengthen the loop of string.*

And this means that the shortest path from F to F′ *via* the tangent at P must be *via* the point P itself.

Finally, then, in something of a masterstroke, we can appeal to Heron's theorem on the reflection of light (Chapter 9) to tell us that PF and PF' must make equal angles with the tangent at P.

Yet, while all this was known in ancient times, no one then can surely have foreseen the extraordinary way in which the ellipse would reappear, over 1500 years later…

Kepler and planetary motion

In 1609, after a painstaking analysis of the astronomical observations, Johannes Kepler proposed the following:

1. The orbit of each planet is an ellipse, with the Sun at one focus.

2. A line drawn from the Sun to a planet sweeps out equal areas in equal times.

The first law, then, is about the shape of the orbit, and the second about the variation in speed as a planet goes round its orbit, moving faster when close to the Sun and slower when further out.

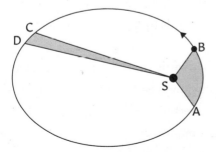

Fig. 91 Kepler's equal-area law.

In Fig. 91, for instance, the two shaded regions have the same area, and—according to Kepler—the planet takes as long to get from C to D as it does to get from A to B.

These were revolutionary ideas, and, it has to be said, not readily accepted at the time.

Did the planets *really* move in this particular way?

And, if so...*why?*

Newton, dynamics, and geometry

In around 1679, Isaac Newton showed that Kepler's second, area-sweeping, law could be explained at once if the planet is acted on by a gravitational force that is directed always *towards the Sun.*

His whole approach is highly geometrical, and, most interestingly, Newton makes extensive use of *the very first theorem in Euclid that really surprised him*, about triangles of quite different shape having equal area (see p. 38).

To see how he does this, we need to note first that Newton does not treat a continuously acting force directly, but replaces it by a succession of short sharp impulses towards S, such as F in Fig. 92.

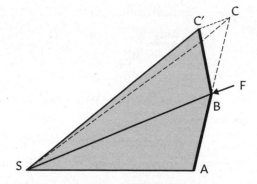

Fig. 92 Geometry in Newton's dynamics.

Consider first the motion when there is no such force. The planet will then—like any other object—move at constant velocity in a straight line, first from A to B, and then, after an equal time interval, from B to C, with AB = BC. The triangles SAB and SBC will therefore have *equal area*, because they have equal 'bases' AB and BC, and the same 'height' (the perpendicular distance from S to AC).

Now consider the effect of the impulse F at B. It produces an instantaneous change in the velocity, in such a way that the planet ends up at C' instead of C, where CC' is proportional to the impulse and in the same direction, i.e. parallel to SB.

It follows, then, that the triangles SBC and SBC' *also* have equal area, because they have the same base SB and are between the same two parallel lines, so have the same height.

So the shaded areas SAB and SBC' must themselves be equal, and in this way Kepler's second law can be explained by assuming a force on each planet directed at all times towards the Sun.

Elliptical motion

It was somewhat later, in 1684, that Newton showed that Kepler's *first* law—elliptical orbit, with the Sun at one focus—implied that the force must be proportional to $1/r^2$, where r denotes distance from the Sun.

This was one of the greatest moments in the history of science, and was decisive in Newton's path towards the theory of universal gravitation.

His whole approach was, again, highly geometrical, involving much sophisticated use of theorems from the world of ancient Greece on conic sections.

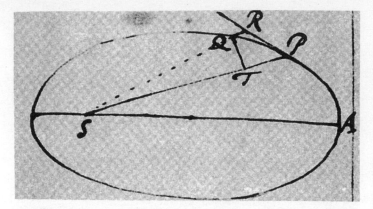

Fig. 93 A sketch of orbital motion from Newton's unpublished manuscript *De Motu corporum in gyrum* (1684). S denotes the Sun.

But what looks, at first sight, like pure Euclidean geometry in Fig. 93, isn't. At one moment the planet is at P, and a short time later it is at Q.

And, crucially, right at the end of his analysis, Newton considers what happens as that time interval gets shorter and shorter, so that Q *gets closer and closer to* P.

This whole kind of thinking propels our geometry story along still further, to very different territory indeed.

And in order to see something of this we must first backtrack a little, by considering a much more algebraic approach to geometry that began to develop in the early seventeenth century.

Geometry by Coordinates

In the early seventeenth century, mathematics took a completely new turn, largely at the hands of Viète, Fermat, and Descartes.

The fundamental idea was to use algebra to help solve problems of geometry, and vice versa.

While the original approaches were rather different, we do this today by first drawing two perpendicular *axes*, and then giving each point a pair of *coordinates* (x, y).

The result of all this is that *a curve can be represented by an equation*, and vice versa.

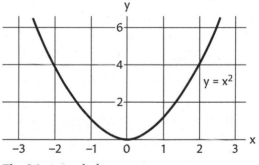

Fig. 94 A parabola.

The equation $y = x^2$, for example, corresponds to a parabola (Fig. 94), which happens to have some remarkable geometrical properties.

In order to see these, however, we will need to grapple with the whole idea of the steepness, or *slope*, of a curve.

And this turns out to be a rather subtle matter, so it makes sense to tackle first the slope of the simplest 'curve' of all: the straight line.

The slope of a line

To get the slope of a line, all we have to do is take two points P and Q on the line, calculate the increases in x and y as we move from P to Q, and divide one by the other (Fig. 95).

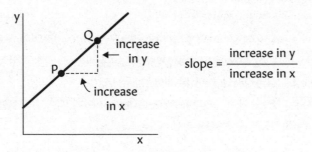

Fig. 95 The slope of a straight line.

Notably, it doesn't matter which two points of the line we choose—this *ratio* is always the same, essentially because of similar triangles, and the fact that the line is straight.

And, fairly evidently, I think, the greater the slope, the steeper the line.

Perpendicular lines

The whole idea of slope is especially useful if two lines are perpendicular, because *the product of their slopes is then* −1.

To see why, imagine turning the line in Fig. 96a anti-clockwise through 90°, as in Fig. 96b.

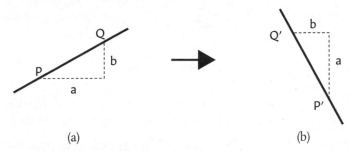

(a)   (b)

Fig. 96 Perpendicular lines.

The original slope is b/a, but as we move from Q′ to P′ in Fig. 96b x increases by b but y *decreases* by a, giving a slope of $-a/b$. So the product of the two slopes is -1.

And the converse is also true: if the product of the slopes is -1, then the two lines are perpendicular.

The distance between two points

If two points have coordinates (x_1, y_1) and (x_2, y_2), the distance between them is

$$D = \sqrt{(x_2 - x_1)^2 + (y_2 - y_1)^2},$$

simply because of the right-angled triangle in Fig. 97.

And determining distances by using coordinates in this way is one of the most common 'practical' applications of Pythagoras' theorem.

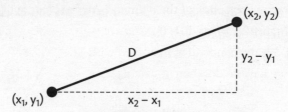

Fig. 97 Using Pythagoras' theorem to find the distance D between two points.

It is also just what we need to see how Thales' theorem emerges from this new, and very different, way of doing geometry...

Thales' theorem (again!)

Our first task, evidently, is to find the equation of a circle, and it is simplest, perhaps, to consider a circle of radius 1 centred on the origin of coordinates $(0, 0)$.

As the distance of each point (x, y) on the circle from $(0, 0)$ must be 1, it follows at once, from what we have just seen, that the equation of the circle must be $x^2 + y^2 = 1$.

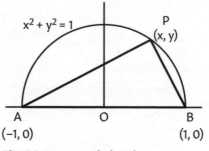

Fig. 98 Proving Thales' theorem.

Moreover, the ends of the diameter AB in Fig. 98 will have coordinates $(-1, 0)$ and $(1, 0)$. And we want to show that AP and BP are perpendicular, in accord with Thales' theorem.

All we have to do, then, is calculate

$$\text{slope of AP} = \frac{y-0}{x-(-1)} = \frac{y}{x+1}$$

$$\text{slope of PB} = \frac{0-y}{1-x} = \frac{y}{x-1}$$

and form the product, which is $\dfrac{y^2}{(x^2-1)}$.

But, as we have seen, $x^2 + y^2 = 1$, because P lies on the circle. So the product of the slopes is -1, and therefore $\angle APB = 90°$.

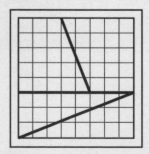

The masterpiece "*8 × 8 grid with 3 lines on it*" has been 'rearranged' by vandals …

Yet nothing has been stolen …

In fact, it's actually BIGGER
than it was before …
13 × 5 = 65 !

How could this happen?
(Answer on p.247)

Geometry and Calculus

His name is Mr. Newton; a fellow of our College, & very young…but of an extraordinary genius and proficiency in these things.

Isaac Barrow, of Trinity College Cambridge,
in a letter of 1669

Calculus is the key to much of modern science and engineering, largely because it is all about the rates at which things change.

But this idea is, itself, closely related to a purely geometric one: *the slope of a curve*.

Given a curve, then, how *can* we determine its slope, or steepness, at some particular point P?

Well, we can presumably get a good approximation to it by choosing some nearby point Q on the curve, and then calculating the slope of the straight line PQ (Fig. 99).

But if we really want the slope of the curve *at* P, it turns out that we have to do something rather more subtle.

In short, we have to explore what happens to the slope of PQ *as Q gets closer and closer to* P.

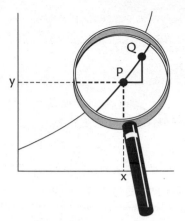

Fig. 99 Finding the slope of a curve.

The slope of a curve

If we actually carry out this procedure with the curve $y = x^2$, for example, it turns out that the slope at any point is $2x$ (see Fig. 100 and Notes, p. 248).

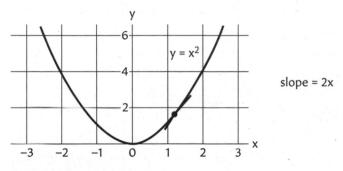

Fig. 100 Slope and tangent.

In this particular case, then, the slope increases with x, which surely makes sense, because the curve $y = x^2$ evidently 'bends upward', and so gets steeper as x increases from 0.

And one reason why the slope of a curve is important is that it is also the slope of the *tangent* to the curve at the point in question.

The curve $y = x^2$, for instance, is actually a parabola, and the slope of the tangent is key to proving one extraordinary property: with a parabolic *mirror*, parallel rays of light from a distant source all get reflected to a single focal point F (see Fig. 101 and Notes, p. 249).

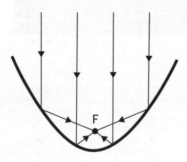

Fig. 101 A parabolic mirror.

Calculus

The whole procedure of obtaining the slope of a curve from its equation is called *differentiation*, and this is the first key idea in calculus.

The subject came fully to life in the second half of the seventeenth century, largely through the work of Isaac Newton, in England, and Gottfried Leibniz, in Germany (Fig. 102).

One major application is to problems where we are trying to find the maximum or minimum value of some quantity of

Fig. 102 (a) Isaac Newton (1642–1727). (b) Gottfried Leibniz (1646–1716).

interest. In Fig. 103, for instance, the slope is positive for small values of x, but negative for larger ones, and one way of determining the maximum value of y, evidently, would be to calculate the value of x for which the slope is actually *zero*.

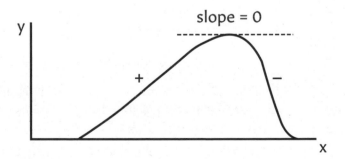

Fig. 103 Using *slope* to find a maximum value.

But one of the greatest achievements of both Newton and Leibniz was to link the whole process of differentiation with two other, apparently quite different, geometrical ideas...

Area and volume

Finding the area or volume of a region enclosed by curved boundaries has always been difficult.

Yet, quite independently, Newton and Leibniz discovered that this can be done, in principle, by 'undoing' or 'reversing' differentiation—a procedure called *integration*.

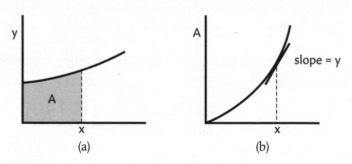

Fig. 104 The fundamental theorem of calculus.

To see something of this, suppose we have a curve, with y depending on x in some way, as in Fig. 104a. Then the area A underneath the curve will also depend on x, and if we plot *that* against x in Fig. 104b we find something that is, at first sight, extraordinary: the *slope* of this new curve, at any particular value of x, is given simply by the corresponding value of y from Fig. 104a (see also Notes, p. 250).

So, if we know how y depends on x, we can, in principle, find how the area A depends on x by reversing the differentiation process.

And similar ideas can be used to find volumes. In 1643, for instance, Evangelista Torricelli caused something of a stir by discovering (by other methods) a three-dimensional object that had *infinite extent but finite volume* (Fig. 105).

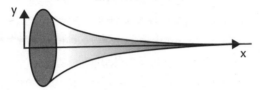

Fig. 105 Torricelli's trumpet, obtained by rotating the curve $y = \frac{1}{x}$, for $x \geq 1$, about the x-axis.

When Thomas Hobbes got to hear of this, some 30 years later, he wrote

> to understand this for sense, it is not required that a man should be a geometrician or a logician, but that he should be mad.

But Torricelli was right, and calculus can be used to confirm it.

More generally, calculus brought about a revolution in mathematics that would eventually take geometry in wholly new directions.

Yet, even as all this was happening, the classical Euclidean geometry of straight lines and circles was still proceeding apace.

And, already, some authors were even trying to 'popularize' it...

A Royal Road to Geometry?

'There is no royal road to geometry.'
Euclid's reply to King Ptolemy I, when asked if there were
some easier way of learning the subject.

Actually, there *is* a *Royal Road to Geometry*; it was a book by Thomas Malton, published in 1774 (Fig. 106).

Malton was a self-taught mathematician who originally kept an upholsterer's shop in London. He also gave private tuition at his house in Poland Street, where he claimed he could

> make any Gentleman (having a Talent for it) a Proficient in
> Geometry, in less than half the Time usually spent in it;
> making it, at the same time, instructive and entertaining.

In his book, Malton is quite forthright in his criticism of Euclid's *Elements*, which he describes as being

> much encumbered with useless Demonstrations.

By this, he appears to mean that there is far too much proving the obvious, which, in Malton's view, serves only to

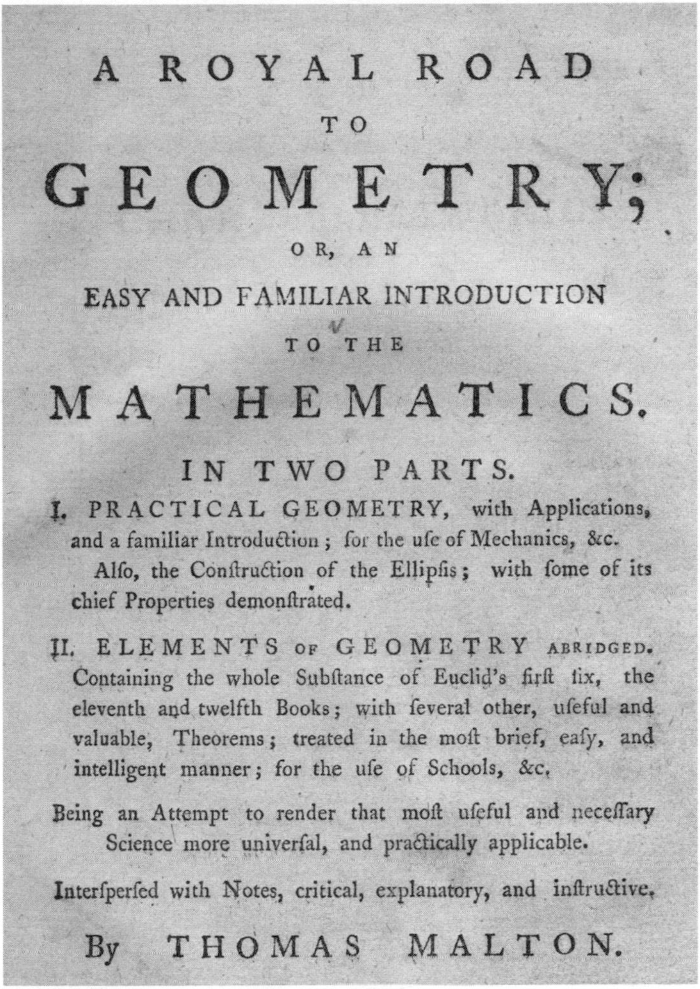

A ROYAL ROAD

TO

GEOMETRY;

OR, AN

EASY AND FAMILIAR INTRODUCTION

TO THE

MATHEMATICS.

IN TWO PARTS.

I. PRACTICAL GEOMETRY, with Applications, and a familiar Introduction; for the ufe of Mechanics, &c.
Alfo, the Conftruction of the Ellipfis; with fome of its chief Properties demonftrated.

II. ELEMENTS OF GEOMETRY ABRIDGED. Containing the whole Subftance of Euclid's firft fix, the eleventh and twelfth Books; with feveral other, ufeful and valuable, Theorems; treated in the moft brief, eafy, and intelligent manner; for the ufe of Schools, &c.

Being an Attempt to render that moft ufeful and neceffary Science more univerfal, and practically applicable.

Interfperfed with Notes, critical, explanatory, and inftructive,

By THOMAS MALTON.

Fig. 106 A royal road to geometry?

perplex and embarrass the minds of Youths.

And he adds

> I have always found more difficulty in demonstrating, to
> another Person, self-evident Propositions, than the most
> intricate of others…

I will return to Malton's book later in the chapter, but let us
take a brief look, first, at two even earlier attempts to stream-
line geometry for the wider public.

The pathway to knowledge?

The first geometry book to be published in English was *The
Pathway to Knowledge*, by Robert Recorde, in 1551.

Recorde is best known, perhaps, as inventor of the 'equals'
sign =, but he was also, by all accounts, one of the greatest
mathematics teachers of all time.

And I have always found one particular observation in the
Pathway rather haunting:

> …it is not easie for a man that shall travaile in a straunge
> arte, to understand at the beginninge bothe the thing that is
> taught and also the juste reason whie it is so…

Accordingly, there are plenty of assertions in the *Pathway*,
but hardly anything by way of proof or deductive argument.

In the case of Pythagoras' theorem, for instance, Recorde
simply draws a diagram for the 3-4-5 case (Fig. 107) and writes

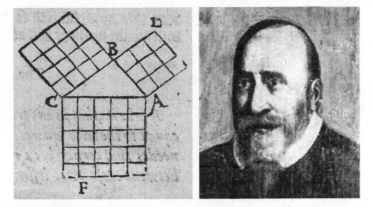

Fig. 107 Robert Recorde, and a figure from his *Pathway to Knowledge* of 1551.

> ...by the number of the divisions in eche of these squares,
>   may you perceave...that the theoreme is true...

But this diagram—sometimes presented as a 'proof' even today—proves nothing, even in the 3-4-5 case, unless we can show somehow that the 50 little squares in it are all exactly the same size.

In any event, Recorde clearly took the view that, for the complete beginner, it was best to focus on the propositions themselves, and not the proofs.

'Plain and easie'

A very different point of view comes over strongly in John Ward's *Young Mathematician's Guide* of 1707 (Fig. 108).

And the author seems to have possessed a certain amount of modesty, for he describes the *Guide* as 'Plain and Easie' and

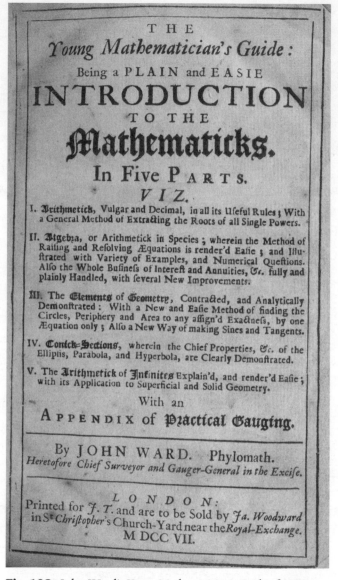

THE
Young Mathematician's Guide:
Being a PLAIN and EASIE
INTRODUCTION
TO THE
Mathematicks.
In Five PARTS.
VIZ.

I. **Arithmetick**, Vulgar and Decimal, in all its Useful Rules; With a General Method of Extracting the Roots of all Single Powers.

II. **Algebra**, or Arithmetick in Species; wherein the Method of Raising and Resolving Æquations is render'd Easie; and Illustrated with Variety of Examples, and Numerical Questions. Also the Whole Business of Interest and Annuities, &c. fully and plainly Handled, with several New Improvements.

III. The **Elements** of **Geometry**, Contracted, and Analytically Demonstrated: With a New and Easie Method of finding the Circles, Periphery and Area to any assign'd Exactness, by one Æquation only; Also a New Way of making Sines and Tangents.

IV. **Conick-Sections**, wherein the Chief Properties, &c. of the Ellipsis, Parabola, and Hyperbola, are Clearly Demonstrated.

V. The **Arithmetick** of **Infinites** Explain'd, and render'd Easie; with its Application to Superficial and Solid Geometry.

With an
APPENDIX of **Practical Gauging**.

By JOHN WARD. Phylomath.
Heretofore Chief Surveyor and Gauger-General in the Excise.

LONDON:
Printed for *J. T.* and are to be Sold by *Ja. Woodward* in St *Christopher's* Church-Yard near the *Royal-Exchange*.
M DCC VII.

Fig. 108 John Ward's *Young Mathematician's Guide* of 1707.

wholly intended to Instruct, and not to amuse or Puzzle the young Learner with hard Words; nor is it my Ambitious Desire of being thought more Learned or Knowing than really I am...

It is, of course, his treatment of geometry that concerns us here, and while everything is clearly influenced by Euclid, Ward frequently goes his own way.

In particular, his proof of the angle-sum of a triangle is different from both ours and Euclid's, and appeals to two corresponding angles and one opposite one (Fig. 109).

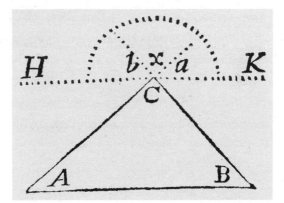

Fig. 109 Proving the angle-sum of a triangle.

But the really distinctive aspect of Ward's treatment of geometry is that as soon as he has the preliminaries out of the way, *he goes off like a rocket.*

For, in the space of just seven pages, he gets through isosceles triangles, ideas of area, Pythagoras' theorem, similar triangles (*and* their connection with area), circle theorems, and more besides, all accompanied by proof.

Quite what Robert Recorde would have made of all this we don't know.

But we do know that *The Young Mathematician's Guide* was one of the most popular and best-selling mathematics books of its time.

Not quite the 'pizza theorem'

I should like to end this chapter by returning to Thomas Malton's *Royal Road to Geometry*.

For a most distinctive feature of that book which sets it apart from several similar ones is that it contains a number of interesting results and theorems that are not in Euclid.

These include, for instance, what we now know as Varignon's theorem, and a little-known theorem which Malton describes as an 'extraordinary property of the Circle'.

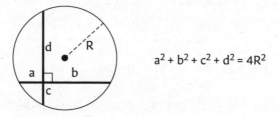

$$a^2 + b^2 + c^2 + d^2 = 4R^2$$

Fig. 110 An 'extraordinary property of the Circle'.

This theorem concerns any two chords which intersect *at right angles* (Fig. 110). And, somewhat remarkably, the sum $a^2 + b^2 + c^2 + d^2$ is then a constant, independent of where the two chords intersect.

More curiously still, some 200 years later, in the 1960s, this result was coupled with some elementary calculus to produce the so-called *pizza theorem* (Fig. 111).

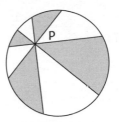

Fig. 111 The pizza theorem.

This gives an exotic way of sharing pizza equally, because, for any internal point P, the total grey area and the total white area are *the same*, provided the angles between the four cuts are all 45°!

I imagine that Thomas Malton cannot possibly have known this, but sense from his 1774 book that he did know that there was still a great deal of interesting and surprising geometry yet to be discovered, even in the thoroughly classical world of circles and straight lines.

Unexpected Meetings

Some of the biggest surprises in geometry come about when lots of straight lines meet unexpectedly at a single point.

This happens, for instance, if we take any old triangle, of no particular shape, and draw a line through each corner perpendicular to the opposite side (Fig. 112).

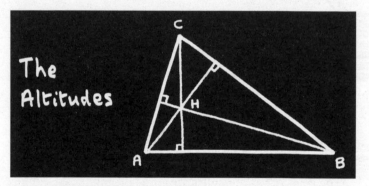

Fig. 112 The altitudes of a triangle.

The way these *altitudes* all meet at a single point has always struck me as a bit peculiar. (It even happens if the triangle ABC has one angle greater than 90°—though the meeting point is then outside the triangle.)

And the history of this result is a little odd, too. It appears to be classical, but is not in Euclid's *Elements*, and the most well-known proof, due to Gauss, dates only from the nine-teenth century.

In any event, in order to understand Gauss' proof we need to begin with a related result which I have always found rather less surprising, and which *is* in Euclid...

The perpendicular bisectors

The perpendicular bisectors of the sides of any triangle ABC meet at a single point, O (Fig. 113).

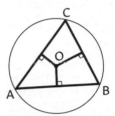

Fig. 113 The perpendicular bisectors.

To see why, imagine, if you will, all the circles that pass through A and B. Only one of them will pass through the third corner C, giving the triangle ABC a unique *circumcircle*, with centre O.

The sides of the triangle are chords of this circle, and it is intuitively clear, I think, by symmetry, that their perpendicular bisectors will all pass through the centre O, and therefore meet at a single point.

A more formal proof is possible, using congruent triangles (see Notes, p. 252). And the meeting point is, once again, *outside* the triangle if one of its angles is greater than 90°.

The altitudes

To prove that the *altitudes* of a triangle ABC meet at a single point, Gauss introduces what appears, at first sight, to be a rather mind-boggling construction. He draws a line through each corner parallel to the opposite side, so as to embed the original triangle ABC in a larger one, DEF (Fig. 114).

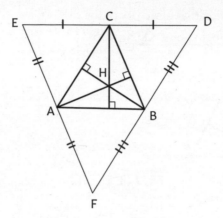

Fig. 114 Gauss' proof that the altitudes meet at a single point.

This creates pairs of parallelograms such as ABCE and ABDC. And as opposite sides of parallelograms are equal, this implies that EC = CD (because both are equal to AB), so that C is the mid-point of ED.

In the same way, A and B are also mid-points, and suddenly we see that the altitudes of triangle ABC *are the perpendicular bisectors of triangle DEF*, and therefore meet at a single point, called the *orthocentre* of the original triangle ABC, and denoted by H.

The angle-bisectors

The three angle-bisectors of a triangle meet at a single point I, called the *incentre* (Fig. 115).

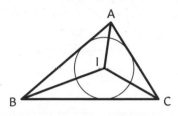

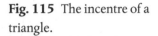

Fig. 115 The incentre of a triangle.

This is evident, I think, *if* we regard it as intuitively obvious that

(a) there is a unique *incircle* within the triangle touching all three sides and

(b) the angle-bisectors will, by symmetry, all pass through its centre I.

Once again, a more formal proof is possible (see Notes, p. 252).

The medians

Take any old triangle, and draw a line from each corner to the mid-point of the opposite side.

Then these 'medians' *also* meet at a single point G, called the *centroid* (Fig. 116).

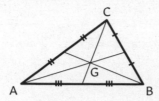

Fig. 116 The medians of a triangle.

To prove it, consider first just the medians through A and B (Fig. 117). Triangles ECD and ACB are similar by SAS, with a 'scale factor' of 2. So ED is parallel to AB, and AB = 2 × ED.

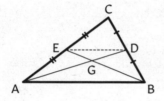

Fig. 117 Proof that the medians meet at a point.

Then, after identifying some alternate angles, we see that triangles EGD and BGA are similar by AA, again with a scale factor of 2.

So

$$AG = 2 \times DG \text{ and } BG = 2 \times EG.$$

In other words, these two medians cut each other at a point which is, in each case, *two-thirds of the way from the triangle corner to the opposite mid-point.*

Applying the same argument to a different pair of medians, we find that all three medians cut each other in this particular way, and therefore meet at a single point.

And there's a lot more…

In a general triangle, of no particular shape

1. The circumcentre O, centroid G, and orthocentre H all *lie on a straight line,* called the *Euler line* (Fig. 118a).

2. The altitudes are the angle-bisectors of the so-called *pedal triangle* in Fig. 118b.

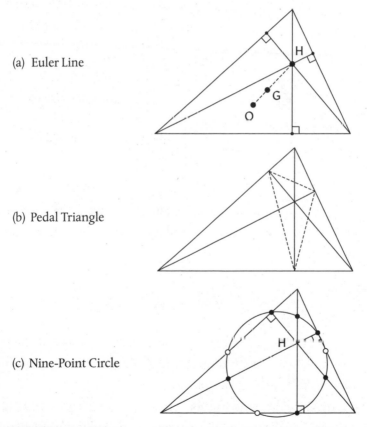

(a) Euler Line

(b) Pedal Triangle

(c) Nine-Point Circle

Fig. 118 Surprising properties of a triangle. (Further details can be found in Notes, p. 253.)

3. The circle through the feet of the altitudes passes through the mid-points of the sides *and* the mid-points of the lines joining H to the corners of the triangle (Fig. 118c).

And the story doesn't end there.

We have looked at just four special points in a triangle, and some of the associated properties. Yet there are many other special points in a triangle, and a famous website currently lists over 16,000 of them (see p. 266).

We will exercise rather more restraint, and introduce just one more in the next chapter, which is largely about a new, and very different, perspective on the ones we have met already.

Ceva's Theorem

This theorem has a very classical feel to it, but was in fact only discovered in 1678.

It concerns what happens when we join each corner of a triangle to some point on the opposite side (Fig. 119).

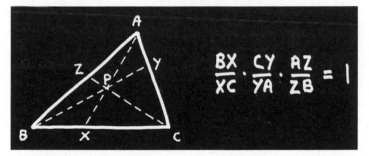

$$\frac{BX}{XC} \cdot \frac{CY}{YA} \cdot \frac{AZ}{ZB} = 1$$

Fig. 119 Ceva's theorem.

Each such point divides its side in a certain ratio, and the theorem itself says that if the three lines meet at a single point P, then *the product of those three ratios must be* 1.

And to prove it we need just one very simple idea.

The triangles A and B in Fig. 120 have the same height, and as the area of a triangle is '$\frac{1}{2}$ base × height', their areas must therefore be *in proportion to their bases*.

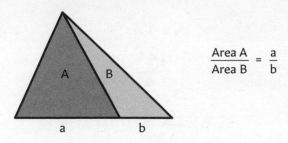

Fig. 120 An area-ratio theorem.

So, to prove Ceva's theorem we begin by observing in Fig. 121 that

$$\frac{\text{Area ABX}}{\text{BX}} = \frac{\text{Area ACX}}{\text{XC}}$$

and

$$\frac{\text{Area PBX}}{\text{BX}} = \frac{\text{Area PCX}}{\text{XC}}.$$

Subtracting, we get the following:

$$\frac{\text{Area ABP}}{\text{BX}} = \frac{\text{Area ACP}}{\text{XC}},$$

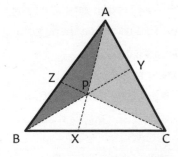

Fig. 121 Proof of Ceva's theorem.

so

$$\frac{BX}{XC} = \frac{\text{Area } ABP}{\text{Area } ACP}.$$

In the same way, then,

$$\frac{CY}{YA} = \frac{\text{Area } BCP}{\text{Area } ABP}$$

and

$$\frac{AZ}{ZB} = \frac{\text{Area } ACP}{\text{Area } BCP}.$$

And when we multiply these three ratios together, *everything cancels* (!), and the product is therefore 1.

Arguably, however, the *converse* of Ceva's theorem is even more important…

The converse of Ceva's theorem

This says that if, in Fig. 122,

$$\frac{BX}{XC} \cdot \frac{CY}{YA} \cdot \frac{AZ}{ZB} = 1,$$

then the three lines AX, BY, CZ meet at a single point.

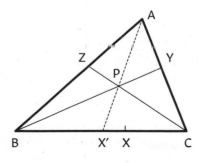

Fig. 122 Proving the converse.

To prove it, let P be the point at which BY and CZ meet, and let AP meet BC at some point X'.

Then by Ceva's theorem itself:

$$\frac{BX'}{X'C} \cdot \frac{CY}{YA} \cdot \frac{AZ}{ZB} = 1,$$

and on combining this with the given condition on X we have

$$\frac{BX'}{X'C} = \frac{BX}{XC},$$

which means that X' and X divide BC in the same ratio. So X' must, in fact, be the point X, which proves the result.

The medians revisited

If X, Y, Z are the *mid-points* of the sides of the triangle, then

$$\frac{BX}{XC} \cdot \frac{CY}{YA} \cdot \frac{AZ}{ZB} = 1,$$

because each individual ratio is equal to 1 (!).

This provides a new 'angle', then, on why the medians meet at a point.

The altitudes revisited

The converse of Ceva's theorem provides a new perspective, too, on why the altitudes meet at a single point.

For, if we draw in one of them (Fig. 123), we see that

$$\frac{BX}{XC} = \frac{c \cos B}{b \cos C},$$

and if we continue round the triangle in this way then Ceva's product is

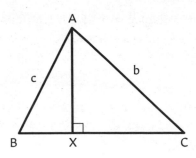

Fig. 123 An altitude.

$$\frac{c\cos B}{b\cos C} \cdot \frac{a\cos C}{c\cos A} \cdot \frac{b\cos A}{a\cos B} = 1.$$

The Gergonne point

A more exotic application arises if we draw the incircle to our triangle and let X, Y, Z be the *points of tangency* (Fig. 124).

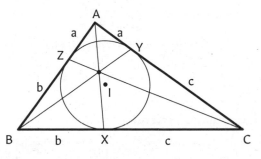

Fig. 124 The Gergonne point of a triangle.

As the two tangents to a circle from an external point are equal, we find that

$$\frac{BX}{XC}\frac{CY}{YA}\frac{AZ}{ZB} = \frac{b}{c}\frac{c}{a}\frac{a}{b} = 1$$

so that AX, BY, and CZ meet at a point.

The point in question is named after J. D. Gergonne (1771–1859), and is not typically the same as the centre of the incircle, I.

How Ceva did it

Ceva was a hydraulic engineer, and discovered his theorem, apparently, using mechanics.

In particular, Archimedes' law of the lever concerns the balance, under gravity, of two masses m and M at different distances from a pivot (Fig. 125).

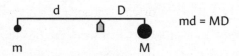

Fig. 125 Archimedes' law of the lever. (The rod joining the two masses is assumed to be weightless.)

Now consider the triangle in Fig. 126, and imagine that we make it out of weightless material and place masses m_A, m_B, and m_C at the corners.

Suppose, too, that we decide to choose those masses so that the point P is the *centre of gravity*, in which case the triangle, when arranged horizontally, will balance on a pinhead at P.

It will then certainly balance on any of the three lines through P in Fig. 126, and three applications of the law of the lever give

$$m_B a_1 = m_C a_2, \quad m_C b_1 = m_A b_2, \quad m_A c_1 = m_B c_2.$$

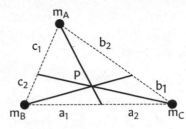

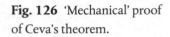

Fig. 126 'Mechanical' proof of Ceva's theorem.

On eliminating the three masses we obtain

$$a_1 b_1 c_1 = a_2 b_2 c_2,$$

which is Ceva's theorem.

And, as it happens, this whole idea of using mechanics to do geometry was not as original as one might suppose.

For we now know that Archimedes himself used methods of this general kind to discover some of his most famous results, including the formula for the volume of a sphere.

Some futher slices of π

1674 **Leibniz** uses **calculus** to show that

$$\frac{\pi}{4} = 1 - \frac{1}{3} + \frac{1}{5} - \frac{1}{7} + \cdots$$

1706 **The symbol π** appears in print for the first time with its modern meaning, in William Jones' *A New Introduction to the Mathematics.*

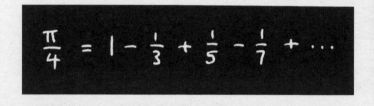

in the *Circle*, the Diameter is to Circumference as 1 to

$$\frac{16}{5} - \frac{4}{239} - \frac{1}{3}\,\frac{16}{5^3} - \frac{4}{239^3} + \frac{1}{5}\,\frac{16}{5^5} - \frac{4}{239^5} -, \&c =$$

3·14159, *&c.* = π. This *Series* (among others for the same purpose, and drawn from the same Principle) I re-

1748 A discovery by **Leonhard Euler** links π with **e** = 2.718 . . . and the **imaginary number** $i = \sqrt{-1}$

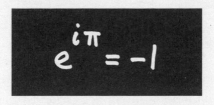

$$e^{i\pi} = -1$$

- -

1767 Johann Lambert proves that π is irrational:

$$\pi \neq \frac{p}{q}$$ where **p** and **q** are whole numbers

A Kind of Symmetry

There is a striking formula, usually credited to Heron of Alexandria, for the area of a triangle in terms of the lengths of its sides (Fig. 127).

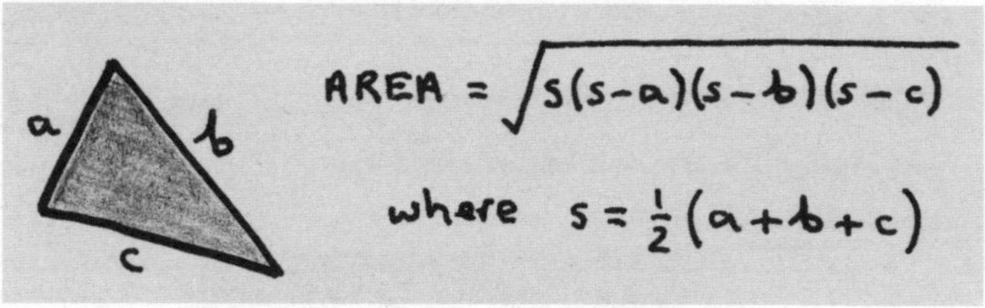

$$\text{AREA} = \sqrt{s(s-a)(s-b)(s-c)}$$

$$\text{where} \quad s = \tfrac{1}{2}(a+b+c)$$

Fig. 127 Heron's formula for the area of a triangle.

It is striking, surely, because of how the three lengths a, b, and c all enter the formula in exactly the same way. Yet a moment's thought reveals this to be rather reassuring.

Suppose, for instance, we have a triangle with sides 5, 6, and 7, and want to know its area. We could set $a = 5$, $b = 6$, $c = 7$ and work it out.

But we would surely be alarmed if setting, say, $a = 7$, $b = 6$, $c = 5$ were to lead to a different answer, because (by SSS) these define essentially the same triangle, but 'viewed' in a different way.

In other words, as the a, b, and c in the formula are just 'the sides of the triangle', with nothing to distinguish which is which, we need to be able to interchange them freely and still get the same answer.

This kind of symmetry often provides a valuable check on an answer, and can also lend a certain elegance to mathematical results.

In particular, the radii of the incircle and circumcircle of a triangle can be written in symmetric ways in terms of the triangle's sides and its area Δ (see Fig. 128 and Notes, p. 258).

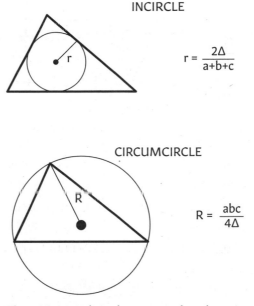

INCIRCLE

$$r = \frac{2\Delta}{a+b+c}$$

CIRCUMCIRCLE

$$R = \frac{abc}{4\Delta}$$

Fig. 128 Incircle and circumcircle radii.

Newton and the altitudes

Symmetry ideas of this kind can also be used to good effect in *proofs*.

A fine example, in my opinion, is a little-known proof by Isaac Newton that the altitudes of a triangle meet at a single point, which appears in an unpublished manuscript of 1680 (Fig. 129).

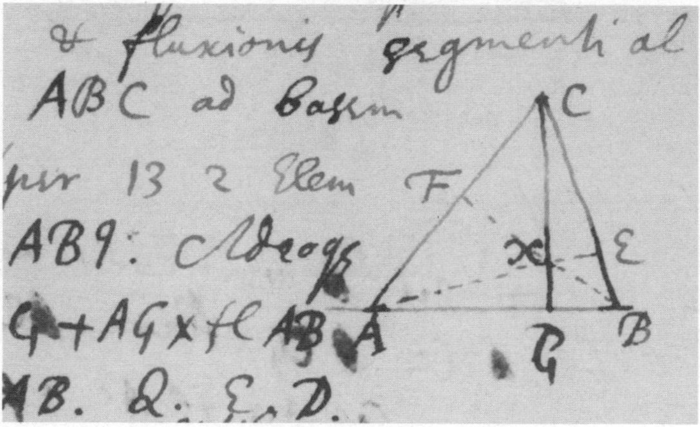

Fig. 129 The diagram for Newton's proof, from his *Geometria curvilinea and Fluxions*, Ms Add. 3963, p54r.

His plan is simply to show, by direct calculation, that the altitudes from A and B meet the one from C *at the same height*.

In Fig. 130, then, we begin by drawing the altitude CD, and denote the three lengths shown by *a*, *b*, and *L*.

And as soon as we draw the altitude through A we find that the marked angles are equal, because they are both $90° - \angle ABC$.

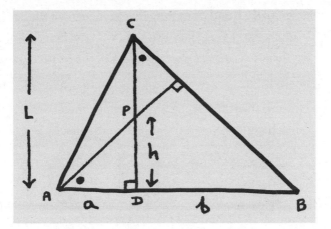

Fig. 130 Two altitudes of a triangle.

So the triangles ADP and CDB are similar, and therefore

$$\frac{h}{b} = \frac{a}{L} \, ,$$

i.e.

$$h = \frac{ab}{L} \, .$$

But this is—unexpectedly, I think—*symmetric in a and b*, so that when we repeat the whole calculation for the altitude from B, all that will happen is that a and b will change places in the formula, L will stay the same, and we will end up with the same value for h as before.

The eyeball theorem

Much the same thing happens in a curious theorem discovered in the 1960s by the Peruvian geometer Antonio Gutierrez.

Take two circles, and draw tangents to each from the centre of the other (Fig. 131). Then the two straight lines AB and A′B′ are always *equal in length*, regardless of the relative size of the two circles.

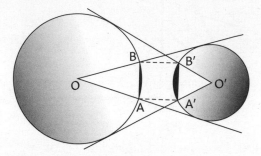

Fig. 131 The eyeball theorem.

Yet the reason becomes clear as soon as we calculate, say, AB in terms of the radii r, $r′$, and the distance D between the centres, for this turns out to be

$$AB = \frac{2rr′}{D},$$

which is unexpectedly symmetric in r and $r′$ (see Notes, p. 259).

The medians, by coordinate geometry

Yet another neat use of symmetry is a coordinate–geometry proof that the medians of a triangle meet at a point.

We need, first, one key result that I have not yet mentioned. If, in Fig. 132, the points P_1 and P_2 have coordinates (x_1, y_1) and (x_2, y_2), then the coordinates of any point P on the line between them can be written

$$x = (1-\lambda)x_1 + \lambda x_2$$
$$y = (1-\lambda)y_1 + \lambda y_2,$$

where λ denotes the fraction P_1P/P_1P_2.

Fig. 132 A point on a line.

Thus, $\lambda = 0$ corresponds to P_1, $\lambda = 1$ to P_2, and the *mid-point*, with $\lambda = \frac{1}{2}$, has coordinates

$$\frac{1}{2}(x_1 + x_2), \frac{1}{2}(y_1 + y_2).$$

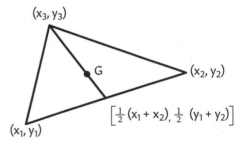

Fig. 133 The medians; a coordinate–geometry approach.

And if we apply this same idea to the median in Fig. 133, we find that the coordinates of a point *two-thirds* of the way from the corner to the mid-point of the opposite side are

$$\frac{1}{3}(x_1 + x_2 + x_3), \frac{1}{3}(y_1 + y_2 + y_3).$$

But these coordinates are completely symmetric with respect to the three corners of the triangle, so, by the same argument, this particular point must lie two-thirds of the way down both the other two medians as well.

In short, and with the help of a little symmetry, we have proved that the three medians of a triangle meet at a single point by considering in detail just *one* of them!

'Pyracy' in Woolwich?

In 1747, Thomas Simpson, a mathematician at the Royal Military Academy in Woolwich (Fig. 134), published a book called *Elements of Plane Geometry*.

And, most unfortunately, a colleague at the academy immediately accused him of plagiarism.

Fig. 134 The Royal Military Academy.

This colleague, the Professor of Fortification and Artillery, claimed that Simpson's book was merely an 'incorrect copy'

of his own work, adding

> The Editor imagined, I suppose, that changing some
> propositions, and mangling the demonstrations of others,
> was a sufficient disguise to make it pass for his own
> performance; but how far this will justify such a piece of
> pyracy, must be left to the judgment of the publick.

For Simpson (Fig. 135) this was, perhaps, just one more
upset in a somewhat turbulent life. For while he eventually
became a Fellow of the Royal Society, he had little formal
education, and taught himself basic mathematics from a
book that he acquired from a fortune-teller. He also married
a widow almost old enough to be his grandmother, who (so it
is said) treated him so badly in later life that he was driven to
'guzzle porter and gin in low company'.

Fig. 135 Thomas Simpson (1710–1761).

In any event, Simpson fought back against the charge of
plagiarism, in a much-expanded second edition of his book,

OF THE

MAXIMA and MINIMA

OF

Geometrical Quantities.

THEOREM I.

If from two given points A, B, *on the same side of an indefinite line* PQ *(in the same plane with them) two lines* AE, BE *be drawn to meet on, and make equal angles* AEQ, BEP *with the said line* PQ; *the lines so drawn, taken together, shall be less than any other two* AG, BG, *drawn from the same points to meet on the same line* PQ.

For, let BNM be per-
pendicular to PNQ, and
let AE be produced to
meet it in M, alfo let
MG be drawn.

Then the triangles
MNE, BNE, having
the angle MEN (=
AEQ[a]) = BEN[b], MNE
= BNE[c], and NE com-
mon to both; have alfo MN = BN, and ME =
BE

[a] 3. 1.
[b] Hyp.
[c] Conftr.

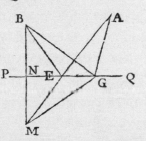

Fig. 136 From the second edition of Thomas Simpson's book, titled *Elements of Geometry* (1760).

published in 1760. And time does not seem to have soothed his anger, for he claimed there to have difficulty giving his side of the story

> without trangressing the bounds of decency.

While Simpson won the argument, the whole episode was something of a shame, because his book really does stand out from many other geometry books of the time.

It starts boldly, with a veiled criticism of Euclid:

> My design, in this little Book, is to lay, before the young Beginner, an easy Method for acquiring a competent Knowledge in the Subject of Geometry, without...being obliged to go through a Number of useless and tedious Propositions.

And he then proceeds quite rapidly, in his own way, with a square-within-a-square proof of Pythagoras' theorem, and, more strikingly still, Varignon's theorem (which was quite new at the time) as early as page 29.

But what really makes Simpson's book distinctive is a later, and most unusual, section on *The Maxima and Minima of Geometrical Quantities*.

What's the smallest area?

Suppose, for instance, that we have two fixed lines AB and AC, and a fixed point E somewhere between them (Fig. 137a).

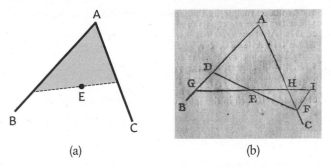

(a)   (b)

Fig. 137 A minimization problem, from Thomas Simpson's
Elements of Plane Geometry (1747).

It is evident, I think, that by choosing the line through E
almost parallel to AB or AC we can make the shaded area as
big as we like.

But how should we choose the line through E to make the
shaded area as *small* as possible?

The answer, in short, is that we need to choose the line so
that E is at its mid-point.

Simpson's proof is illustrated in Fig.137b, where DE = EF,
and GEH is some other line through E. By introducing FI, parallel to GD, he creates two congruent triangles EDG and EFI.
As a result, the area of EDG is greater than the area EFH, so by
changing from the line DEF to the line GEH we gain more
'shaded' area than we lose. The line DEF must therefore be the
one enclosing least area.

And while this solution is surely an elegant one, I would
argue that the next problem from Simpson's collection has a
more elegant solution still, not least because it features, at a

crucial moment, an ingenious appeal to the converse of Thales' theorem...

Queen Dido's problem

According to legend, Queen Dido arrived at the North African coast in about 800 BC, and was granted as much land there as she could encompass with an ox-hide.

She promptly cut the ox-hide into very thin strips, and stitched them all into one very long strip, but this still left her with a problem.

Fig. 138 Dido's problem (with a straight coastline).

With reference to Fig. 138, the problem is to maximize the shaded area by varying the points A, B, and the shape of the curve C, while keeping the *length* of the curve C fixed.

And Queen Dido—so it is said—knew the answer: C should be a semi-circle, with AB as diameter.

Now, if we pass over the question of whether a maximum actually *exists* at all (which can be a subtle matter in problems of this general kind), then Simpson presents an elegant proof that the answer must, indeed, be a semicircle.

Observe, first, that the area of a triangle with two given sides will be a maximum when those sides are perpendicular (Fig. 139), as a direct consequence of the formula $\frac{1}{2}$ base × height.

Fig. 139 Which has the largest area?

Next, imagine that we have found the arrangement giving the maximum area in Queen Dido's problem (Fig. 140a), and that there is some point D on the enclosing curve such that ∠ADB is *not* 90°.

We can then quickly obtain a contradiction by changing the length of AB until ∠ADB *is* 90°, while keeping both the size and shape of the segments ACD and DEB unchanged (Fig. 140b).

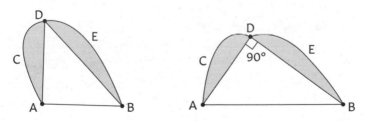

Fig.140 The heart of Simpson's argument.

For, if you would have the area ACDEBA, contained by some right-line AB, and ACDEB whereof the length is given, to be the greatest possible, and ADB, at the same time, not a right-angle: Then, let PSQ be a right-angle, contained under PS = AD, and QS = BD; and, having joined

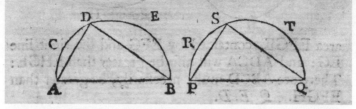

206   *Of the Maxima and Minima*

PQ, upon PS and QS conceive two figures PRS and QST to be formed, equal, and alike in all respects to ACD and DBE.   Since the area PSQ is greater than ADB [c]; it is manifest, that the area PRSTQP, contained by the right-line (PQ) and PRSTQ (= ACDEB [d]) will also be greater than the area ACDEBA [e], *which is repugnant :* Therefore the area ACDEBA cannot be the greatest possible, unless the angle ADB be a right one. *Q. E. D.*

•Theor.4.

• Hyp.

• Ax. 6.

COROLLARY.

Hence, because the angle in a semi-circle is a right-angle [f], it is evident that the area will be the greatest possible, when the given length, or boundary, forms the arch of a semi-circle ; whereof the indetermined right-line proposed is the diameter.

' 13. 3.

Fig. 141 Simpson's argument, as it appears in his *Elements of Geometry* of 1760. The result is presented as a corollary to his Theorem 14. The reference to '13.3' is to Thales' theorem.

The length of the curve ACDEB will be unchanged, but we will have increased the enclosed area, contrary to hypothesis, because the new triangle ADB will have greater area than the old one, while the other two areas will be the same as before.

It follows, then, that $\angle ADB$ must be 90° for *all* points D on the maximizing curve.

As a final flourish, we appeal to the *converse of Thales' theorem*, and conclude that the maximizing curve must be a semicircle with AB as diameter.

Interestingly, Simpson appears to slip up at this very last step by appealing to Thales' theorem itself, rather than its converse.

More curiously still, the whole ingenious argument seems to be universally credited, in all the books and papers that I have seen, to the renowned Swiss geometer Jakob Steiner, in the 1840s.

So far as I can determine, however, it was really invented by Simpson, almost 100 years earlier.

Fermat's Problem

One of the most famous minimization problems in geometry is this:

> *Given a triangle ABC, find the point P which*
> *makes* PA + PB + PC *as small as possible.*

It dates from 1644, when Fermat posed it as a challenge to Torricelli, who was working in Florence.

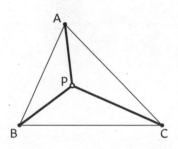

Fig. 142 The Fermat problem.

Torricelli solved it, and with one proviso, the solution is really rather neat: choose P so that *all three angles at* P *are* 120° (Fig. 142).

He did observe, however, that this only works if all the angles of the original triangle ABC are less than 120°.

Torricelli's approach

His main argument can be loosely paraphrased as follows.

Let P be the desired minimizing point, or *Fermat point*, in Fig. 143, and draw the ellipse with foci B and C which passes through P.

Imagine, now, moving round the ellipse. As the sum of the distances from B and C will be constant (by the property shown in Fig. 89), and P minimizes PA + PB + PC, it follows that P must be the point on the ellipse that is *nearest to* A.

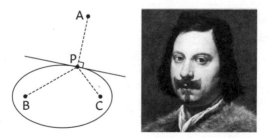

Fig. 143 Torricelli's main argument, as inferred from his *Opere*, vol. 1, first published in 1919.

This means, then, that PA is perpendicular to the tangent at P—as in Fig. 143. Moreover, PB and PC make equal angles with that tangent (by the 'reflection property' of Fig. 90), which implies that ∠APB = ∠APC.

Repeating the argument with, say, A and B as the foci then gives the result—all three angles at P must be equal.

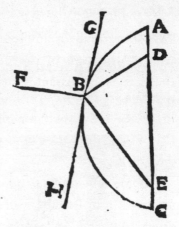

ESt enim, F B, omnium minima, quæ ad rectã, GH, duci poſſunt, & ſubindè minima quoque erit omnium ad ſemiellipſim, ABC, ab, F, ducibilium. Inſuper anguli, F B G, F B H, ſunt æquales, quia recti; & anguli, G B D, H B E, pariter inter ſe æquales. Ergo &, F B D, F B E, erunt æquales.

Fig. 144 Cavalieri seeking the Fermat point of triangle ACF, in his *Exercitationes Geometricae Sex* of 1647.

Three years later, in 1647, Bonaventura Cavalieri, published a very similar solution, which he had apparently obtained independently (Fig. 144). Cavalieri also observed that if one corner of the triangle happens to contain an angle greater than 120° we need to do something rather more mundane, and choose P as that wide-angled corner.

Then, a little later still, one of Torricelli's students, Vincenzo Viviani, found a different approach altogether...

Viviani's approach

This elegant solution to the Fermat problem uses an elementary theorem due to Viviani himself.

Viviani's theorem

This states that for any internal point P of an *equilateral* triangle the sum of the perpendicular distances from the sides is a constant, independent of the position of P (Fig. 145).

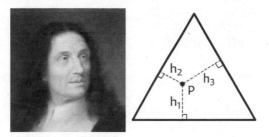

Fig. 145 Viviani's theorem: in an equilateral triangle, $h_1 + h_2 + h_3 = $ constant.

And the proof is very simple: just join P to the three corners, dividing the triangle into three smaller ones. Then if the original triangle has side L and height H we have, from areas,

$$\frac{1}{2}(h_1 + h_2 + h_3)L = \frac{1}{2}HL,$$

which proves the theorem and shows that the constant value of $h_1 + h_2 + h_3$ is, in fact, the height H of the equilateral triangle.

This gives us what we need, then, for an entirely different attack on the Fermat problem.

The Fermat problem (again)!

In Fig. 146, let PA, PB, and PC make equal angles of 120° with each other. We want to show that this minimizes the sum PA + PB + PC.

To do this, draw in the dashed lines *perpendicular* to PA, PB, PC. As the angles of any quadrilateral add up to 360°, the angles of the triangle DEF must all be 60°, so that triangle must therefore be *equilateral*.

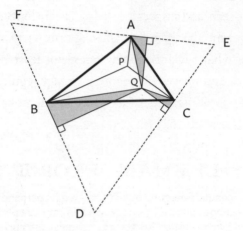

Fig. 146 An alternative approach to the Fermat problem.

Now consider any other internal point Q of the triangle ABC.

Plainly, from Fig. 146, the sum QA + QB + QC will be greater than the sum of Q's *perpendicular* distances from the sides of triangle DEF, which will, itself, be equal to $PA + PB + PC$, by Viviani's theorem.

So the 'three-equal-angles' point P does indeed minimize the sum of the distances to the corners of triangle ABC; any other internal point Q leads to a greater value.

While it is often claimed that this whole approach is due to Torricelli, and simply *published* by Viviani in his *De Maximis et Minimis* of 1659, I have found no evidence for this in the sources that I have seen, and suspect it is due to Viviani himself.

In any event, the treatment above is only a *very* loose paraphrase of what Viviani actually does, not least because he announces and proves his theorem for *any* regular polygon (i.e. one with all its sides and angles equal), not just for an equilateral triangle (Fig. 147).

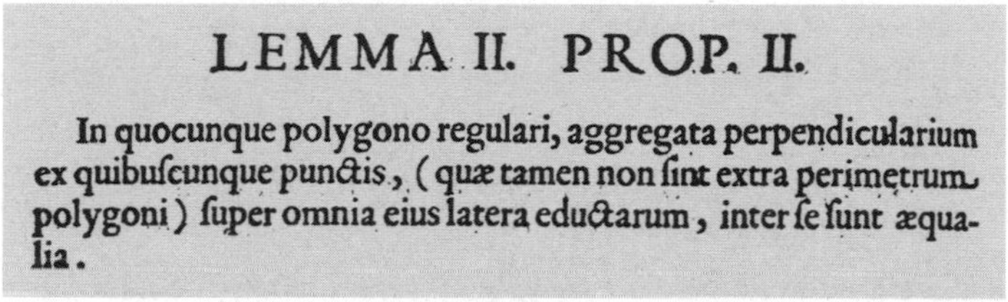

LEMMA II. PROP. II.

In quocunque polygono regulari, aggregata perpendicularium ex quibuſcunque punctis, (quæ tamen non ſint extra perimetrum polygoni) ſuper omnia eius latera eductarum, inter ſe ſunt æqualia.

Fig. 147 Viviani's theorem, as it appears in his *De Maximis et Minimis* (1659).

Thomas Simpson's approach

Not surprisingly, Simpson includes this problem in his collection of 1747, and tackles it in his own way.

He begins by recalling Heron's result on the reflection of light at a plane mirror, which we discussed in Chapter 9.

Next, he shows that a similar result holds for two points outside a *circle* with centre O: the shortest path between them, via a point P on the circle, is such that the two parts of the path make equal angles with the radius OP.

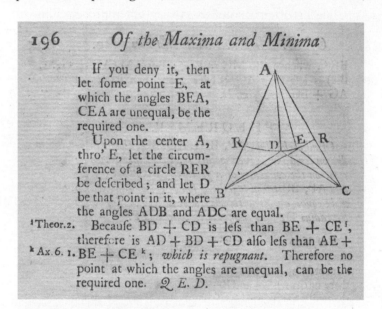

196   *Of the Maxima and Minima*

If you deny it, then let some point E, at which the angles BEA, CEA are unequal, be the required one.

Upon the center A, thro' E, let the circumference of a circle RER be described; and let D be that point in it, where the angles ADB and ADC are equal.

[Theor.2.] Becaufe BD + CD is lefs than BE + CE[i], therefore is AD + BD + CD alfo lefs than AE + [Ax.6.1.] BE + CE[k]; *which is repugnant.* Therefore no point at which the angles are unequal, can be the required one. *Q. E. D.*

Fig. 148   From Simpson's *Elements of Geometry* (1760).

He then applies this to the Fermat problem, establishing that all three angles must be 120° in a proof by contradiction (Fig. 148).

In a separate publication, *The Doctrine and Application of Fluxions* (1750), he notes—like Cavalieri before him—the quite different solution if one angle of the triangle happens to be greater than 120°.

Constructing the Fermat point

As part of his original solution to the problem, Torricelli gave a practical way of actually constructing the Fermat point P.

On each side of the original triangle ABC, construct an equilateral triangle and its circumcircle (Fig. 149a). The three circumcircles will then meet *at the Fermat point* P.

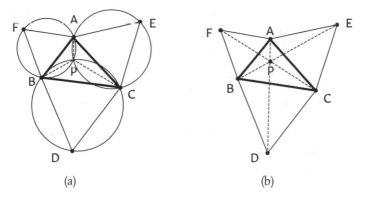

(a)   (b)

Fig. 149 Constructing the Fermat point.

To see why this happens, let P denote the Fermat point, so that $\angle BPC = 120°$. As $\angle BDC = 60°$, the four points P, C, D, B all lie on a circle, by the *converse* of the circle theorem on p. 70. In other words, the circumcircle of triangle BCD passes through P, and the same argument applies, of course, to the other two circles as well.

As it happens, however, when we come to the actual construction process, we don't really need to bother with the circles at all.

This is because our three equilateral triangles, together with the second circle theorem on p. 68, imply that all six angles around P in Fig.149b are 60°!

This in turn means that APD, BPE, and CPF are all *straight lines*. So all we have to do to locate P is construct, say, the top two equilateral triangles in Fig. 149b, draw in the lines BE and CF, and see where they meet.

A 'proof by rotation'

I would like to end this chapter with a particularly elegant solution to the Fermat problem.

We begin by choosing any point P inside the triangle ABC, and *rotate the triangle PAB about B through* 60° (Fig. 150).

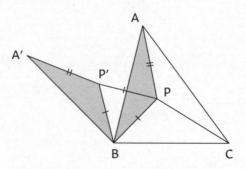

Fig. 150 Solving Fermat's problem 'by rotation'.

This creates an equilateral triangle BPP', and, in consequence, the original minimization problem turns out to be equivalent to minimizing the length of the path CPP'A'.

Now, C and A' are *fixed* points, but P is 'movable', and P' will also move in response to how we choose P.

And if we choose P so that PA, PB, and PC all make angles of 120° with each other, the path CPP'A' becomes *straight* (because $\angle BPP' = \angle BP'P = 60°$, while $\angle BPC = \angle BP'A' - 120°$), and therefore the shortest possible.

While this proof is usually credited to J. E. Hofmann, in 1929, many of the ideas underlying it are really much older—especially when we notice that ABA', too, is an equilateral triangle, sitting on one side of the original, and that the minimizing straight line CPP'A' from Fig. 150 is none other than the straight line CPF in Fig. 149b.

A *Soap Solution*

Sometimes, with a really difficult minimization problem, *physics* can lend a helping hand.

Fig. 151 A soap film approach to geometry.

Suppose, for instance, we have two Perspex plates separated by four pins at the corners of a square, and we dip all this into a bowl of water with a bit of washing-up liquid.

When we take it out again, there will be some kind of soap film linking the pins.

Moreover, if we keep on doing this, then, sooner or later, a particularly distinctive configuration will emerge, with five straight portions and 2 three-way intersections (Fig. 151).

And this is, in fact, the solution to a tricky geometrical problem.

What's the shortest network?

Imagine, if you will, that our job is to link four towns by a road network which is *as short as possible*.

And suppose, too, that the towns lie—somewhat conveniently—at the corners of a square of side 1.

Then two possibilities which come quickly to mind, I think, are shown in Figs. 152a and 152b. They certainly allow someone at any particular town to travel to any other town, which is our basic requirement.

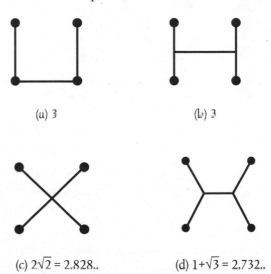

(a) 3

(b) 3

(c) $2\sqrt{2} = 2.828..$

(d) $1+\sqrt{3} = 2.732..$

Fig. 152 In search of the shortest network.

With a little thought, however, we soon find that we can do a bit better by using the diagonals (Fig. 152c).

Yet we do better still with the network shown in Fig. 152d, with two three-way junctions where the roads make 120° angles with each other.

To see this, observe that we now have some 30°, 60°, 90° triangles (Fig. 153).

So $QB \sin 60° = \frac{1}{2}$, and as $\sin 60° = \sqrt{3}/2$ we find that

$$QB = \frac{1}{\sqrt{3}}.$$

Also, $QS = QB \cos 60°$, and as $\cos 60° = 1/2$ we have

$$QS = \frac{1}{2\sqrt{3}}.$$

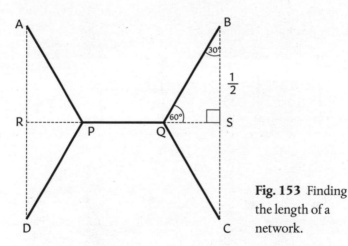

Fig. 153 Finding the length of a network.

Finally, as RS = 1, the middle segment PQ will be $1 - 2 \times QS$, i.e. $1 - 1/\sqrt{3}$. So, if we add this to the four parts like QB we end up with

$$1 + \sqrt{3} = 2.732\ldots$$

as our grand total for the whole network, which is certainly the shortest so far.

Now, as it happens, this is *the shortest of all*, and therefore the solution to our problem, but proving this is not easy.

And, unfortunately, our soap film experiment doesn't 'prove' it, either. A soap film will always try to minimize its

(a)

(b)

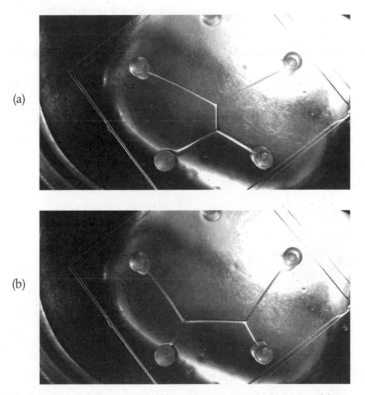

Fig. 154 Two different soap film 'solutions' to the same problem.

surface energy, which is proportional to its surface area, and hence—in our particular case—to the length of the network. But the laws of physics guarantee only that when the soap film finally settles it will have less surface energy than it would have in any *slightly* different state. For all we know, there is some *very substantially* different state with less energy still.

The difficulty can be seen more clearly, perhaps, if our system has slightly less symmetry, and in Fig. 154 the arrangement of the pins is not quite square.

When I first withdrew the apparatus the film settled into state (a), but by gently blowing on the film through a drinking straw I decreased the length of the short middle section until the film suddenly jumped into a completely different state (b), which, as it turns out, is slightly shorter than (a).

It's all a bit like throwing a ball carelessly over some bumpy ground, with lots of hills and valleys (Fig. 155). With a bit of

Fig. 155 A simple system with multiple equilibrium states.

friction present, we know that, sooner or later, the ball will end up at the bottom of one of the valleys, but we have no reason to suppose that it will necessarily end up at the bottom of the deepest one.

Nonetheless, the soap film experiments not only help point the way towards a solution, but offer a fresh perspective on why the three-way junctions here—and indeed in the Fermat problem of Chapter 23—always involve angles of 120°.

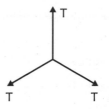

T   T   **Fig. 156** Equal forces in equilibrium.

Viewed in terms of forces, rather than energy, this is simply because the different parts of the soap film pull on each other *equally*, and the only way three equal forces can be in balance at a single point is if the angles involved are all 120° (Fig. 156).

The LADIES *Diary:*

OR, THE

Woman's ALMANACK,

For the YEAR of our LORD, 1740.

Being the *BISSEXTILE,* or LEAP-YEAR:

Containing many Delightful and Entertaining *Particulars,* Peculiarly Adapted for the *Use* and *Diversion* of the

FAIR-SEX.

Being the *Thirty-Seventh* ALMANACK ever Publish'd of this Kind

1. HAIL! happy LADIES of the *BRITISH* Isle,
   On whom the GRACES and the MUSES smile.

2. LONG had your lovely *Shape,* and matchless *Mein,*
   The Wonder of the Neighb'ring Nations been;

3. NATURE to make your *Triumph* morcompleat.
   To peerless CHARMS has added piercing WIT.

4. NO more let *SCYTHIA* vaunt her FEMALE-HOST,
   Nor their SEMIRAMIS th' *Assyrians* boast:
   WIT join'd to BEAUTY, *Fame* shall now record;
   Which lead more Captive than the Conqu'ring Sword.

Printed by *A. Wilde,* for the Company of STATIONERS, 1740.

Fig. 157 *The Ladies Diary* (1740).

Geometry in The Ladies' Diary

The *Ladies' Diary* was a popular journal, published annually in London, from 1704 till 1841.

It was, in part, an almanac, but it also contained

Delightful and Entertaining Particulars

which included some mathematical problems sent in by its readers.

And, as it turned out, it was these problems that gave *The Ladies' Diary* its lasting fame.

Keeping one's head

Many of the contributors were, in fact, men. In the 1715 edition, for instance, we find the following problem, posed by 'Mr. Tho. Shepheard':

If you walk all the way round the Earth's equator,
your head travels further than your feet.
How much further, if your height is *h*?

Actually, this is a very loose paraphrase of the original problem, because (a) the person is 5 foot 7 inches tall, (b) they don't go all the way round, and (c) the whole problem is cast in *verse* so appalling that it is, I believe, best forgotten.

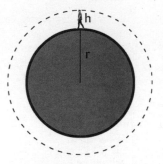

Fig. 158 A walk round the Earth.

In any event, the answer to the version of the problem in Fig. 158 is

$$2\pi h$$

and—somewhat surprisingly, perhaps—*independent of the radius r of the Earth!*

Yet this is simply because the feet travel a distance $2\pi r$, the head a distance $2\pi(r + h)$, and on subtraction the terms involving r cancel.

Pythagoras in the garden

The questions usually required rather more persistence and ingenuity.

In the 1754 edition, for instance, a problem from 'Mr Timothy Doodle' asks for the radius of a circular fountain, given the distances of its rim from the four corners of a rectangular garden. And the published solution (Fig. 159) involves four applications of Pythagoras' theorem, together with a bit of algebra.

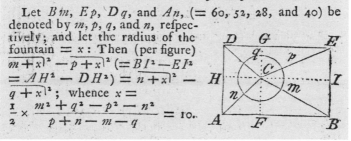

Fig. 159 Lots of Pythagoras, in a problem from *The Ladies Diary* (1754).

Look through any window ...

A slightly offbeat similar triangles problem appeared in the 1790 issue, and was solved by Miss Betty Claxton of Benwell, near Newcastle upon Tyne. Her solution, published the following year, is shown in Fig. 160.

The problem itself is to determine the height of a window CD, given three obscure pieces of information. First, from the point L we can see 3½ yards of the wall of the house opposite. Second, if we move 5 yards closer to the window, to the point S, we can see 6 yards of the wall. Finally, the house opposite is 12 yards beyond the window.

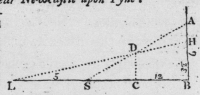

Fig. 160   Betty Claxton's solution. Here *a:b::c:d* means *a/b* = *c/d*.

It is all, essentially, a novel variation on the AD 850 'ladder problem' of Chapter 8.

A penny-farthing problem

This problem, from the 1800 issue, is about proving the formula for the distance *D* (Fig. 161).

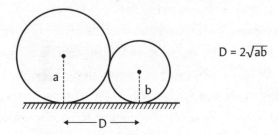

Fig. 161   A penny-farthing problem.

The symmetric way in which the radii a and b appear in the answer is reassuring, for if we 'nip round the back' we have essentially the same problem, with plainly the same answer, except that a and b will have swapped places.

Oddly enough, the whole problem resurfaced 150 years later, in a most peculiar way.

For, in May 1952, Johanna Mankiewicz, 15-year-old niece of the famous film producer, was faced with just this problem as part of her geometry homework in Los Angeles.

And, unable to solve it, she wrote to Albert Einstein.

Fig. 162 A homework hint from Albert Einstein.

Just as remarkably, Einstein replied, possibly spurred on a little by

> I realize that you are a very busy man, but you are the only person we know of who could supply us with the answer…

a remark which did not go down at all well with the principal of Johanna's school, especially when the whole story was splashed all over *The New York Times* and countless other newspapers around the world.

Curiously, Einstein's reply (Fig. 162) doesn't have the two circles touching, and gives, in effect, only a general 'hint' on how to solve the problem.

A tricky problem

In truth, many of the questions in *The Ladies' Diary* were more difficult than the ones I have chosen so far.

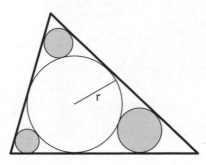

Fig. 163 A tricky problem from *The Ladies Diary* (1730).

One quite early example, from 1730, is shown in Fig. 163, where the problem is to determine the radius *r* of the incircle. Yet we are told nothing directly about the triangle itself, only the radii *a, b, c* of the small circles.

This would seem to be a *much* more difficult problem, and I, at least, have some trouble viewing it as one of the 'Delightful and Entertaining Particulars' trumpeted on the journal's cover (see Notes, p. 260).

Yet the answer itself *is* something of a delight, because it is not only reassuringly symmetric in *a, b,* and *c,* but unexpectedly simple:

$$r = \sqrt{ab} + \sqrt{bc} + \sqrt{ca}.$$

* * *

The Ladies' Diary ceased publication in 1841, and in order to take our story still further we need to recognize that, by this time, mathematicians were beginning to seriously question the very foundations of geometry.

And if we wish to see something of this, the first thing we need to do—not surprisingly—is take a much closer look at what Euclid actually did…

EUCLID 1847

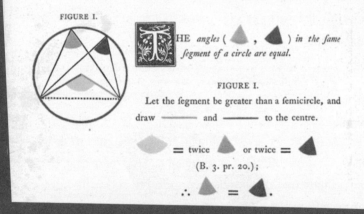

BOOK III. PROP. XXI. THEOR.

FIGURE I.

THE angles (,) in the same segment of a circle are equal.

FIGURE I.

Let the segment be greater than a semicircle, and draw ———— and ———— to the centre.

= twice or twice =
(B. 3. pr. 20.);

∴ = .

Oliver Byrne's extraordinary edition of 1847 claimed that, with **COLOUR**, Euclid's **Elements** could be mastered

"in less than one third the time usually employed".

THE FIRST SIX BOOKS OF

THE ELEMENTS OF EUCLID

IN WHICH COLOURED DIAGRAMS AND SYMBOLS

ARE USED INSTEAD OF LETTERS FOR THE

GREATER EASE OF LEARNERS

BY OLIVER BYRNE

SURVEYOR OF HER MAJESTY'S SETTLEMENTS IN THE FALKLAND ISLANDS
AND AUTHOR OF NUMEROUS MATHEMATICAL WORKS

LONDON
WILLIAM PICKERING
1847

What Euclid Did

Euclid's *Elements* consists of 13 'Books', each containing a logical sequence of propositions. And each proposition is immediately followed by its proof.

Book I deals with many of the fundamentals, including congruent triangles, parallel lines, and area. It ends on a high note, with Pythagoras' theorem.

I	Triangles, parallels, and area
II	Geometric algebra
III	Circles
IV	Constructions
V	Theory of proportions
VI	Similar figures
VII–IX	Number theory
X	Irrationals
XI–XIII	Solid geometry

Fig. 164 The structure of Euclid's *Elements*.

Book II takes a geometric approach to several ideas that we would probably view algebraically, such as that shown in Fig. 165. But algebra—at least as we know it today—is essentially a sixteenth-century invention, so Euclid had to find another way.

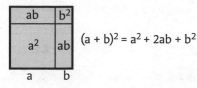

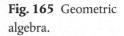

Fig. 165 Geometric algebra.

Book III is about circle geometry, and the early parts of our Chapter 11 follow Euclid's treatment fairly closely.

Euclid deals with similar triangles very late on, in Book VI, the reason being that the whole subject is full of ratios of lengths, and these can all too easily be *irrational* (Fig. 166).

Fig. 166 An irrational number in geometry.

This was a major problem for the ancient Greeks, who simply did not view irrational numbers like $\sqrt{2}$ *as numbers at all*. And to deal with this, Euclid precedes his treatment of similarity with a long, and highly sophisticated, 'theory of proportions', due to Eudoxus, in Book V.

Books VII–X are, for the most part, not directly about geometry at all. In Book VIII, Proposition 20, for instance, we find Euclid's famous proof that there are infinitely many prime numbers.

He returns to geometry in Books XI–XIII, but mostly, now, in three dimensions, culminating in a treatment of the five regular polyhedra, or Platonic solids (Fig. 167).

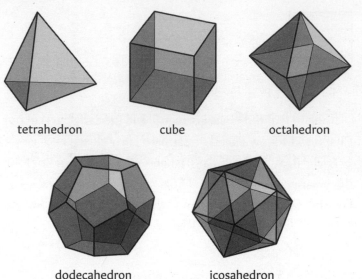

tetrahedron   cube   octahedron

dodecahedron   icosahedron

Fig. 167 The Platonic solids. In each one, all the faces are identical regular polygons.

While some editions (such as Barrow's) include a further two Books, these are not now thought to be by Euclid.

Euclid's postulates

Euclid tries to state all his assumptions *at the very outset*.

He also tries to use as few as possible, and just five assumptions—variously called 'postulates' or 'axioms'—relate specifically to geometry.

The first three are about constructions (Fig. 168), which Euclid often uses to establish the *existence* of certain geometrical concepts that we might, more casually, take for granted.

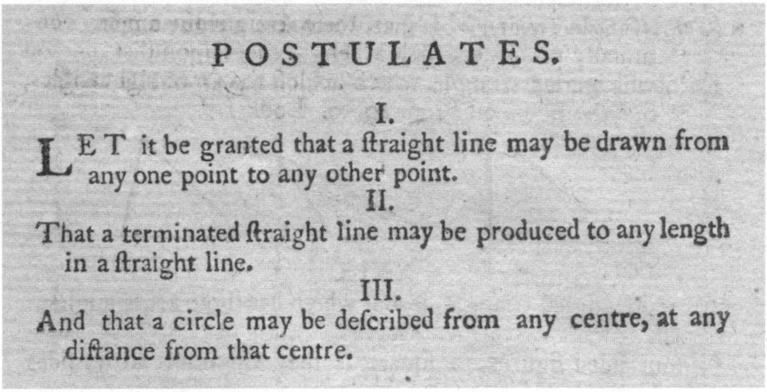

POSTULATES.

I.

LET it be granted that a ſtraight line may be drawn from any one point to any other point.

II.

That a terminated ſtraight line may be produced to any length in a ſtraight line.

III.

And that a circle may be deſcribed from any centre, at any diſtance from that centre.

Fig. 168 Euclid's first three postulates, from an edition of the *Elements* by Robert Simson (1804).

The fourth postulate is rather different:

All right angles are equal to one another

and, in my experience, totally mystifying to a newcomer to geometry, who is liable to think: 'Of course they are! They're all 90 degrees!'

It is now usually interpreted as a statement about the nature of space itself, namely that it is both homogeneous (the same in all places) and isotropic (the same in all directions).

The idea is used, for instance, in the proof that opposite angles are equal (Fig. 169), when we assert that, *despite their different directions*, the lines AB and CD both contain at E an angle of 180°—or, as Euclid would say, two right angles.

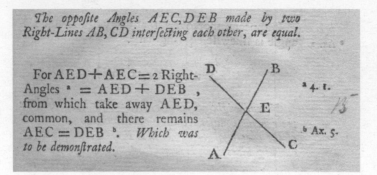

Fig. 169 Proof that opposite angles are equal, in Thomas Simpson's *Elements of Plane Geometry* (1747).

The fifth and final explicit starting assumption is Euclid's famous *Parallel Postulate*:

> If a straight line falling on two straight lines make the
> interior angles on the same side less than two right
> angles, the two straight lines, if produced indefinitely,
> meet on that side on which are the angles less than the
> two right angles.

Needless to say, it is quite unlike the others, and in terms of sheer subtlety it is in a league of its own. For this reason, we will discuss it separately, in Chapter 27.

Interestingly, Euclid tries to advance as far as he can *without* using the parallel postulate in the earlier parts of Book I (Fig. 170), and that is where we turn next.

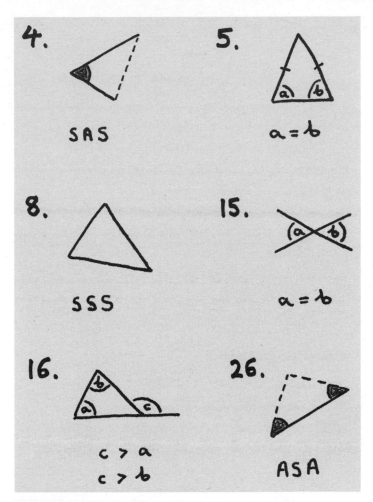

Fig. 170 Some early propositions from Euclid, Book I.

The earlier parts of Book I

Congruence

Euclid starts with SAS, and 'proves' it by arguing that if one triangle is placed on top of the other it will fit exactly (Fig. 171).

Fig. 171 Euclid's 'proof' of SAS congruence.

It is not at all obvious, however, that his fundamental assumptions allow us to do this sort of thing. And, to put it another way, if they *do*, then do we really need the fourth postulate?

It could be argued, then, that SAS is more or less—as in the present book—a starting assumption.

But while we also assumed ASA and SSS congruence as fairly 'obvious', Euclid emphatically does not; he *proves* them (see Notes, p. 261).

To indicate how this kind of thing can be done I will present here a later, and rather elegant, proof of SSS due to Philo of Byzantium (*c*.280–220 BC).

A proof of SSS congruence

Let two triangles satisfy the SSS congruence conditions. We want to prove that they really *are* congruent.

To do this, take one triangle and place it underneath the other, in the specific way shown in Fig. 172, and join CR.

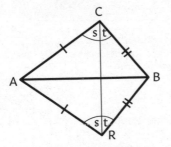

Fig. 172 Philo's proof of SSS congruence.

Now, triangle RAC is isosceles, so the angles s are equal. Triangle RBC is also isosceles, so the angles t are equal. So $\angle ACB = \angle ARB$, and the triangles ACB and ARB are therefore congruent by SAS.

This proof assumes, of course, the principal result about isosceles triangles, and we turn next to how Euclid goes about proving that.

Euclid on isosceles triangles

Euclid's proof that the base angles of an isosceles triangle are equal comes very early in the *Elements*, when all he has at his disposal is congruence by SAS.

And his argument is a sophisticated one that has flummoxed beginners ever since (Fig. 173).

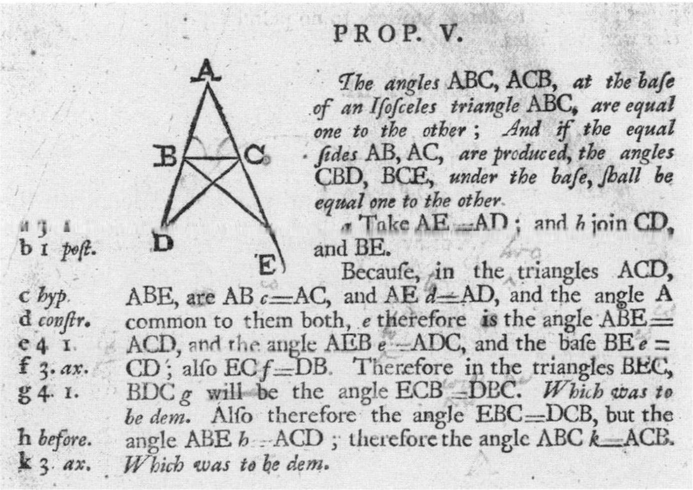

PROP. V.

The angles ABC, ACB, at the base of an Isosceles triangle ABC, are equal one to the other ; And if the equal sides AB, AC, are produced, the angles CBD, BCE, under the base, shall be equal one to the other.

a 3. 1.
b 1 post.   Take AE =AD ; and b join CD, and BE.

c byp.   Because, in the triangles ACD, ABE, are AB c=AC, and AE d=AD, and the angle A
d constr.   common to them both, e therefore is the angle ABE=
e 4 1.   ACD, and the angle AEB e =ADC, and the base BE e =
f 3. ax.   CD ; alfo EC f=DB. Therefore in the triangles BEC,
g 4. 1.   BDC g will be the angle ECB =DBC. Which was to
              be dem. Alfo therefore the angle EBC=DCB, but the
h before.   angle ABE h =ACD ; therefore the angle ABC k=ACB.
k 3. ax.   Which was to be dem.

Fig. 173 Euclid's proof that the base angles of an isosceles triangle are equal, from Barrow's edition.

Why doesn't he present a *simpler* proof?

There is one, for instance, that involves no constructions at all, and is usually credited to Pappus (end of AD 3rd century). We observe, simply, that, in Fig. 173, triangles BAC and CAB are congruent, by SAS, so $\angle ABC = \angle ACB$.

It seems to me, at least, that this is logically valid, even though it is really only a formal version of 'If we nip round the back of an isosceles triangle it'll all look the same, won't it?'

We can be more clear, I think, why Euclid doesn't use the proof I presented in Chapter 4 involving the angle-bisector AD and SAS (Fig. 174).

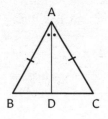

Fig. 174 The angle-bisector approach to isosceles triangles.

The reason, almost certainly, is that Euclid hasn't at this stage proved—from his list of stated assumptions—that such a thing as an angle-bisector *exists*, and this is, again, just one more indication of how demanding the *Elements* can be in terms of logical rigour.

Euclid on Parallel Lines

Euclid *defines* parallel lines as follows:

> Parallel straight lines are straight lines which, being in
> the same plane and being produced indefinitely in both
> directions, do not meet one another in either direction.

The parallel postulate

A slightly informal paraphrase of Euclid's parallel postulate is
given in Fig. 175.

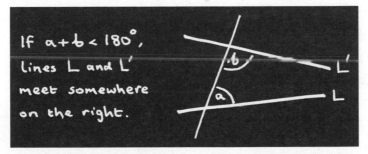

Fig. 175 The parallel postulate (somewhat informally).

Now, there are 180° in a straight line. So if, instead, $a + b$ is
greater than 180° in Fig. 175, the equivalent angle-sum on the

left will be less than 180°, and by the postulate itself the two lines will then meet somewhere on the *left*.

So, according to Euclid's parallel postulate, the only way the two lines can be parallel is if $a + b = 180°$.

Or, to put it slightly differently, if L and L′ are parallel, then $a + b = 180°$. And if we appeal again to 180° in a straight line, it follows that

> *If two lines are parallel, alternate angles are equal* (Fig. 176).

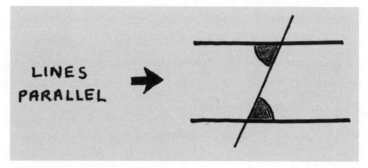

Fig. 176 If two lines are parallel…

Moreover, as opposite angles are equal, this is equivalent to our first major assumption in Chapter 2: if two lines are parallel, the corresponding angles are equal.

But while we immediately went on to assume the *converse*, too, Euclid emphatically does not; he presents—in effect—a *proof* of it, as we see next.

Book I, Proposition 27

This proposition, indicated schematically in Fig. 177, is as follows:

> *If alternate angles are equal, the two lines are parallel.*

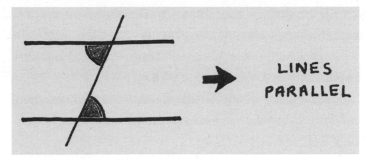

Fig. 177 If alternate angles are equal...

Euclid's proof of this appeals to a somewhat obscure (and, I think, relatively little-known) Proposition I.16, where he presents a proof that the 'exterior' angle of any triangle, *c*, is greater than either of the interior angles *a* and *b* (Fig. 178a).

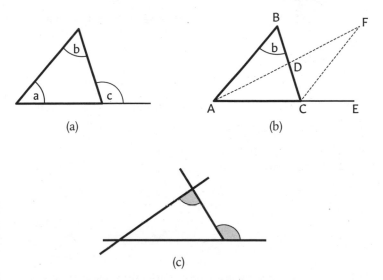

(a)   (b)

(c)

Fig. 178 Diagrams for Euclid I.16 and I.27.

To prove that $b < c$, he joins A to D, the mid-point of BC (Fig.178b), and extends the line by its own length to F. Triangles ADB and FDC are then congruent by SAS, so that $b = \angle DCF$, which is less than $\angle DCE = c$.

He then observes, in I.27, that if alternate angles are equal, and the lines in question were *not* parallel, they would have to meet (Fig. 178c)—giving an immediate contradiction with the result $b < c$ in I.16.

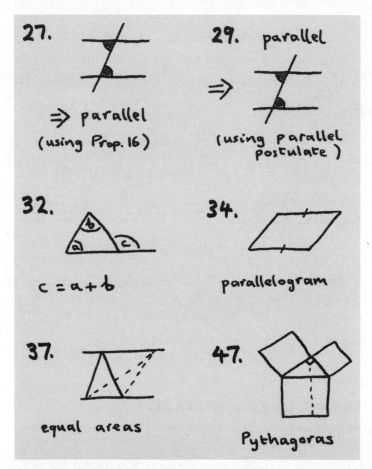

Fig. 179 Some later propositions in Euclid Book I.

The angle-sum of a triangle

This comes quite late in Book I, as Proposition 32, because it depends critically on ideas about parallel lines.

Euclid's proof is slightly different from the one in Chapter 2, and draws on one pair of alternate angles and one pair of corresponding angles.

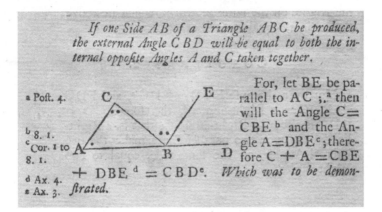

If one Side AB of a Triangle ABC be produced, the external Angle CBD will be equal to both the internal opposite Angles A and C taken together.

For, let BE be parallel to AC ;.[a] then will the Angle C= CBE [b] and the Angle A=DBE[c]; therefore C + A =CBE + DBE [d] = CBD[e]. *Which was to be demonstrated.*

a Post. 4.
b 8. 1.
c Cor. 1 to 8. 1.
d Ax. 4.
e Ax. 3.

Fig. 180 Euclid I.32, as it appears in Thomas Simpson's *Elements of Plane Geometry* (1747).

With reference to Fig. 180, he now proves, essentially, that the exterior angle CBD is not simply greater than either angle A or angle C (Proposition I.16); it is actually equal to their *sum*.

And from this it quickly follows that the three internal angles of the triangle must add up to 180°.

Similar triangles

We saw in Chapter 8 how ideas of similarity are related to those of area, but they are bound up with ideas of parallelism as well, and all these things come together in Euclid's treatment of the subject, in Book VI.

The cornerstone of this treatment is Proposition 2, where he takes a triangle ABC and draws in DE, parallel to BC (Fig. 181).

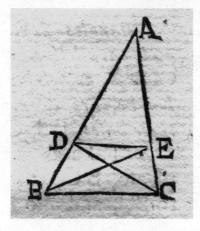

Fig. 181 The figure for Euclid VI.2, from Barrow's edition.

The two triangles ABC and ADE are then similar, and Euclid establishes that DB/AD = EC/AE as follows.

First, by the area-ratio theorem of Chapter 20:

$$\frac{DB}{AD} = \frac{Area\,DBE}{Area\,ADE}$$

and

$$\frac{EC}{AE} = \frac{Area\,ECD}{Area\,ADE}.$$

Finally,

$$Area\,DBE = Area\,ECD,$$

by viewing these two triangles *in a different way*, with the same 'base' DE and standing between the same two parallels, DE and BC.

Euclid's result then follows, and on adding 1 to both sides we find that

$$\frac{AB}{AD} = \frac{AC}{AE},$$

so that *similar triangles have sides which are in the same proportion.*

PROOF by PICTURE?

One proof of Pythagoras' theorem has become quite popular as an **animation**...

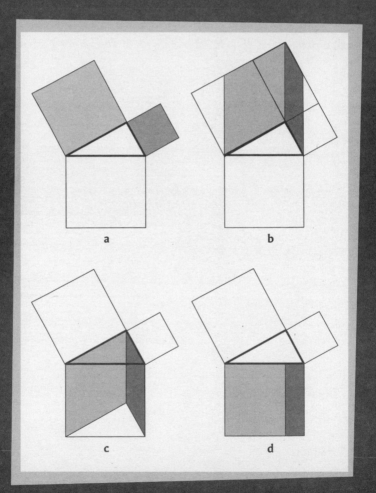

a

b

c

d

This whole idea seems to date from Antonio Lecchi's *Elementa Geometriæ* of **1753**:

If we extend HK and ON to meet at L, triangles AKL and BAC are congruent (by SAS) and at right angles to one another.

So LADF **is** a straight line, LA=BC=DF, and everything in the animation *does* fit!

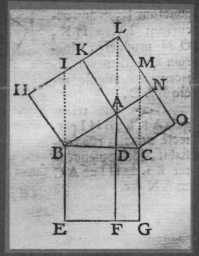

28

'A New Theory of Parallels'?

The year 1888 saw the appearance of a very strange book by the Oxford mathematician C. L. Dodgson—better known as Lewis Carroll, the author of *Alice in Wonderland*.

Its title was *A New Theory of Parallels*, and in the book Dodgson showed that we can replace Euclid's parallel postulate, if we wish, by an obscure assumption involving a hexagon inscribed in a circle (Fig. 182).

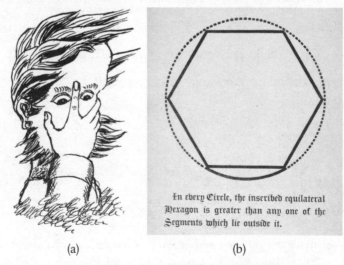

In every Circle, the inscribed equilateral Hexagon is greater than any one of the Segments which lie outside it.

(a)   (b)

Fig. 182 (a) C. L. Dodgson (self-portrait) and (b) his alternative to Euclid's parallel postulate.

The assumption is that the area of the hexagon is greater than the area of any one of the circular segments outside it, and Dodgson himself wrote of this alternative:

> that it is somewhat *bizarre* I am willing to admit.

But then the parallel postulate itself is a bit odd, not least because it seems at first sight to be all about two lines *not* being parallel.

Little wonder, then, that ever since Euclid's time mathematicians have looked around for alternatives.

And it eventually became clear that, given his other assumptions, Euclid's geometry can be developed perfectly satisfactorily using various alternatives to the parallel postulate, including the assumption that

(i) the angle-sum of any triangle is 180°

or that

(ii) triangles can be similar without being congruent.

But the main alternative, still in frequent use today, is not nearly so revolutionary.

In fact, it is only subtly different from the parallel postulate itself…

Playfair's postulate

In its usual modern form, this is:

> Given a straight line L and a point P not on the line, there is one and only one line through P parallel to L (Fig. 183).

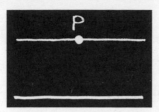

Fig. 183 Playfair's postulate.

While this is commonly attributed to the Scottish mathematician John Playfair, it actually appears rather differently in his *Elements of Geometry* of 1795:

> Two straight lines which intersect one another cannot be both parallel to the same straight line.

In addition, Playfair attributes the whole idea to others, including a book called *The Rudiments of Mathematics* by Ludlam, some ten years earlier (Fig. 184).

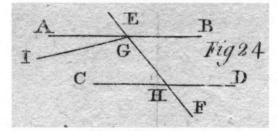

Fig. 184 From Ludlam's *Rudiments of Mathematics* (1785).

In any event, Playfair's book is significant in other ways too. In particular, it contains over 50 pages of 'Notes' on Euclid's *Elements*, including an extensive discussion of parallel lines.

And he observes that one of the great difficulties occurs at the very start, when we *define* parallel lines as lines in the plane which never meet.

It is only natural to ask, then, whether we can avoid some of the difficulties by actually defining parallel lines in an entirely different way...

'Parallel' = 'constant distance apart'?

This is the definition of parallel lines adopted by Thomas Simpson in his *Elements of Plane Geometry* of 1747 (Fig. 185).

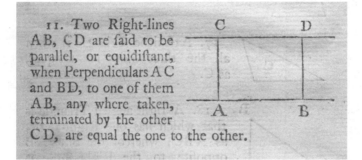

11. Two Right-lines AB, CD are faid to be parallel, or equidiftant, when Perpendiculars A C and B D, to one of them AB, any where taken, terminated by the other CD, are equal the one to the other.

Fig. 185 Thomas Simpson's original (1747) approach to parallel lines.

But, as Playfair points out, there's a difficulty.

The original line is AB, and we are invited to draw perpendiculars, on the same side, all of equal length, from *any* point on the line AB.

But it only takes *two* such perpendiculars to define a straight line 'parallel' to the original, so how do we know that the ends of *all the other* perpendiculars also lie on that line?

In fact, Simpson himself appears to have realized the difficulty, because in the second edition of his book, in 1760, he adopts a considerably more sophisticated approach.

'Parallel' = 'in same direction'?

It is even more tempting, perhaps, to simply define parallel lines as being 'in the same direction'.

And, if asked what we mean by that, exactly, we could say that they *make equal angles with any third line that intersects them both*.

This amounts, really, to taking 'corresponding angles are equal' as our *definition* of parallel lines (Fig. 186).

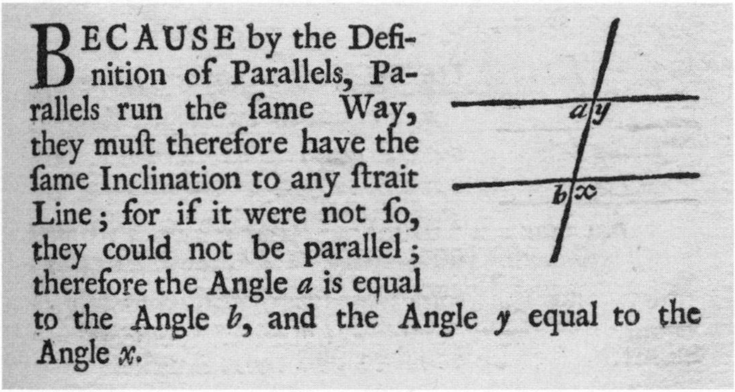

B ECAUSE by the Definition of Parallels, Parallels run the same Way, they muſt therefore have the same Inclination to any ſtrait Line; for if it were not ſo, they could not be parallel; therefore the Angle *a* is equal to the Angle *b*, and the Angle *y* equal to the Angle *x*.

Fig. 186 From John Ward's *Compendium of Speculative Geometry*, published (posthumously) in 1765.

But Playfair again points out a difficulty.

We are entitled, presumably, to construct two lines that make equal angles with one particular third line of our choice, and call them parallel, but how do we then know that these two lines also make equal angles with *any other* third line that crosses them both?

Can we, perhaps, prove it?

A false proof...

Here's how *not* to do it.

In Fig. 187, let the angles a be given as equal. Then from one of the triangles

$$b = 180° - a - d$$

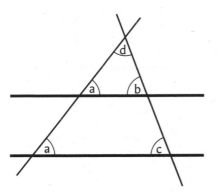

Fig. 187 A false proof...

and from the other

$$c = 180° - a - d$$

and therefore $b = c$.

The trouble with this 'proof', of course, is that we have used the fact that the angle-sum of a triangle is always 180°, and all the proofs of this that we have seen so far depend on the very thing that we are trying to prove.

...and another...

Perhaps we can prove that the angle-sum of a triangle is 180° *without* using ideas of parallelism?

Suppose we just let the angle-sum of a triangle be S, say.

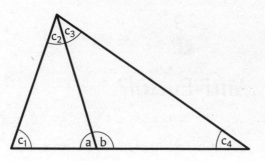

Fig. 188 …and another one…

Then, in Fig. 188,

$$c_1 + c_2 = S - a$$
$$c_3 + c_4 = S - b.$$

If we now add, we find on the left-hand side the angle-sum of the large triangle, so $S = 2S - a - b$. But $a + b = 180°$, and therefore

$$S = 180°.$$

The mistake this time is perhaps a little more subtle, but in the very first line. While we didn't make the classic error of assuming what we were trying to prove—that the angle-sum of a triangle is always $180°$—we *did* assume it is a *constant*, S, across three different triangles.

At best, then, the argument above shows that *if* the angle-sum of a triangle is a constant, then that constant has to be $180°$.

And the only reason I have included both these false proofs at this stage is to point out the kind of pitfalls that lie in wait for the unwary as soon as we try to explore the very foundations of geometry in any depth.

Anti-Euclid?

No one really knows, I think, whether Euclid's *Elements* was ever intended for genuine beginners. But by the second half of the nineteenth century it was certainly being used that way in UK schools, even with quite young pupils.

Fig. 189 J. M. Wilson (1836–1931), an early reformer of geometry teaching in schools.

Schoolteachers themselves were apparently divided on whether this was a good idea. One mathematics master at Eton, for instance, claimed that

the Euclid hour is always a bright one,

while J. M. Wilson (Fig. 189), during his time at Rugby School, wrote of

the extreme repulsiveness of Euclid to almost every boy.

In May 1870, Rawdon Levett, a young schoolmaster in Birmingham, wrote a letter to *Nature* which included the sentence

If the leaders of the agitation for the reform of our geometrical teaching would organise an Anti-Euclid Association, I feel sure they would meet with considerable and daily-increasing support.

And, partly in response to this, the year 1871 saw the foundation of the Association for the Improvement of Geometrical Teaching (Fig. 190), which later became the Mathematical Association.

The A.I.G.T. was never exactly an 'anti-Euclid' body, but it certainly challenged Euclid as the standard introduction to geometry in schools.

This was greatly welcomed by some, but bitterly opposed by others, including Charles Dodgson at Oxford. In *Euclid and his Modern Rivals* (1879) he vigorously defended the status quo, and lampooned the A.I.G.T. mercilessly as

'The Association for the Improvement of Things in General'.

Yet the case against Euclid *for beginners* is substantial.

Association for the Improvement of Geometrical Teaching.

The General Meeting is to be held at University College, Gower Street, W.C., on Saturday, January 14, 1893.

At the Morning Sitting (11 a.m.), the Report of the Council will be read; the new officers will be elected; and the following will be proposed for election as members of the Association:—

> P. J. HEAWOOD, M.A., Durham.
> Prof. G. LORIA, Genoa.
> W. H. WAGSTAFF, M.A., Birmingham.

After the conclusion of the formal business, Mrs. BRYANT, D.Sc., will give **'A Model Lesson on Geometry, as a basis for discussion.'**

After an adjournment for luncheon at 1 p.m., members will reassemble (2 p.m.) to hear papers by Mr. G. HEPPEL, M.A. **('The Use of History in teaching Mathematics'),** and Mr. F. E. MARSHALL, M.A. **('The Teaching of Elementary Arithmetic').**

All interested in the objects of the Association are invited to attend.

Members who wish to have any special matter brought forward at the General Meeting, but who are unable to attend, are requested to communicate with one of the Honorary Secretaries.

[P. T. O.

Fig. 190 An early meeting of the A.I.G.T.

First, making real sense of Euclid's logical structure is extremely demanding, partly because the mathematical backdrop at the time was so different to that of today.

The *Elements* can also be criticized as austere—perhaps unnecessarily so. Here's J. M. Wilson again:

> Euclid places all his theorems and problems on a level,
> without giving prominence to the master-theorems.

Another difficulty is the whole idea of proof by contradiction, or *reductio ad absurdum*. While Euclid uses it a great deal—and seems, at times, to almost relish it—the idea can be conceptually difficult for beginners.

In 1867 this led, in fact, to one of the most bizarre editions of Euclid ever, by Lawrence S. Benson of New York, who had something of an obsession about removing all 'proof by contradiction' and replacing it with something else (Fig. 191). Unfortunately, he also claimed to prove that the area of a circle of radius r is exactly $3r^2$, and Benson's *Geometry* eventually became an object of ridicule, with the *Southern Review* of 1869 reporting that

> No book could, in our humble opinion, be better adapted
> to unhinge all the reasoning powers of the juvenile mind.

In any event, the main trouble with Euclid for beginners, surely, is that there is just so much 'proving the obvious', simply because he tries to assume so little at the outset.

GEOMETRY:

THE ELEMENTS OF EUCLID AND LEGENDRE

SIMPLIFIED AND ARRANGED

TO EXCLUDE FROM GEOMETRICAL REASONING

𝔗𝔥𝔢 𝔕𝔢𝔡𝔲𝔠𝔱𝔦𝔬 𝔄𝔡 𝔄𝔟𝔰𝔲𝔯𝔡𝔲𝔪;

Fig. 191 From the title page of Benson's *Geometry* (1867).

As the mathematician A. N. Whitehead (1861–1947) once observed:

> It requires a very unusual mind to undertake the analysis of the obvious.

And in W. D. Cooley's *Elements of Geometry, Simplified and Explained*, for instance, published in 1860, we find this passage:

> Curiosity is awakened by the discovery of remarkable and unsuspected properties.... These rewards of labour are, in the ordinary system, too long delayed.

To this end, Cooley races along, not quite as fast as John Ward in his *Young Mathematician's Guide*, but in a similar spirit.

And there is another feature, too, that makes his book distinctive. Just occasionally, in what are essentially short 'asides', he mentions entirely informal approaches that help bring the subject to life.

Fig. 192 From W. D. Cooley's *Elements of Geometry, Simplified and Explained* (1860).

In connection with the angle-sum of a triangle for instance, he advocates tearing the corners off a paper triangle and re-assembling them (Fig. 192).

This proves nothing, of course, except that the angle-sum is approximately 180° in one particular case. But it was most unusual for a book at this time to include such an 'experimental' approach alongside the formal development.

The twentieth century

By the first half of the twentieth century, the reforms instigated by the A.I.G.T. (and others) were well under way, and the leading geometry books for schools included elements of 'practical geometry', often before the deductive treatment (Fig. 193).

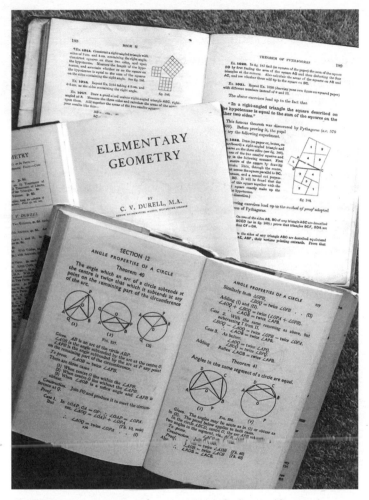

Fig. 193 Geometry, as I first met it. Books by Godfrey and Siddons (1919), Durell (1934) and, bottom, P. Abbott's *Teach Yourself Geometry* (1959).

In 1946, the President of the Mathematical Association observed in his annual lecture that Euclid's *Elements* was, indeed, a peak of Greek culture, but then added:

> How quite it came to be inferred from this that the book was pre-eminently suited for the education of small boys in Victorian England...forms a curious chapter in the history of education in this country.

In any event, whatever we may think of Euclid as an *introduction* to geometry, especially for young people, we surely acknowledge it as the most extraordinary and influential book on mathematics ever written.

And even today, perhaps, some will have a sneaking sympathy with one headmaster of Eton in the mid-nineteenth century, of whom it is said

> He divided the books of the world into three classes:
> Class I: The Bible
> Class II: Euclid
> Class III: All the rest.

When Geometry Goes Wrong...

One of the strangest books on my shelves at home, in Oxford, claims that Pythagoras' theorem is actually *wrong*.

In Fig. 194, for instance, we have (supposedly) a right-angled triangle ABC, with the two shorter sides both of length 1. Yet according to the book in question, the 'true' length of the hypotenuse is not $\sqrt{2}$, but $\frac{3}{2}$.

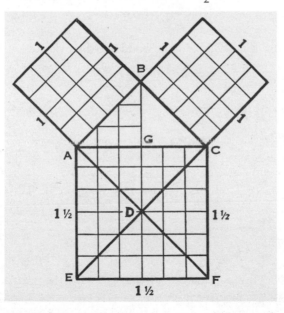

Fig. 194 From Heisel's *Mathematical and Geometrical Demonstrations* (1934).

The author of the book, Carl Theodore Heisel, published it himself in 1934, and it is now something of a collector's item, though sadly for all the wrong reasons.

So far as I can determine, Heisel simply rejects outright the idea that irrational numbers such as $\sqrt{2}$ can have any place in geometry.

Which brings me to his real obsession—and I do mean obsession—the numerical *value of* π...

Pi in the sky

Tucked into my copy of Heisel's book is a typewritten covering letter, signed by Heisel himself, and dated 6 April 1934 (Fig. 195).

In it, he claims to prove that the *exact* value of π is

$$\frac{256}{81} = 3.160...,$$

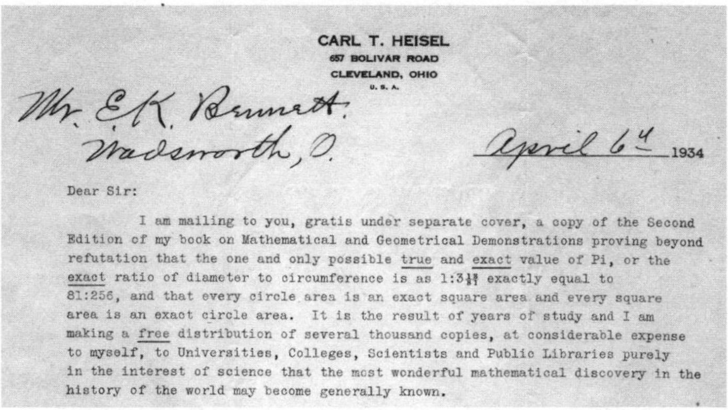

CARL T. HEISEL
657 BOLIVAR ROAD
CLEVELAND, OHIO
U. S. A.

Mr. E. K. Bennett.
Wadsworth, O.   April 6ᵗʰ 1934

Dear Sir:

I am mailing to you, gratis under separate cover, a copy of the Second Edition of my book on Mathematical and Geometrical Demonstrations proving beyond refutation that the one and only possible true and exact value of Pi, or the exact ratio of diameter to circumference is as 1:3¹⁴⁄₈₁ exactly equal to 81:256, and that every circle area is an exact square area and every square area is an exact circle area. It is the result of years of study and I am making a free distribution of several thousand copies, at considerable expense to myself, to Universities, Colleges, Scientists and Public Libraries purely in the interest of science that the most wonderful mathematical discovery in the history of the world may become generally known.

Fig. 195 Covering letter from Heisel to Mr. E. K. Bennett, of Wadsworth, Ohio (1934).

and, casting any modesty aside, describes this as 'the most wonderful mathematical discovery in the history of the world'.

Yet we saw in Chapter 14 how Archimedes proved that π must lie between $3\frac{10}{71}$ and $3\frac{1}{7}$, which corresponds to

$$3\cdot1408\ldots< \pi <3\cdot1428\ldots.$$

So, not content with taking on Pythagoras, Heisel takes on Archimedes as well.

In the interests of balance we could, I suppose, note than an earlier π-enthusiast, James Smith of Liverpool, claimed to prove in 1869 that π is actually

$$\frac{25}{8}=3\cdot125,$$

which, in contrast, is too *small* to be consistent with Archimedes.

Smith inflicted his ideas on various mathematicians of the day, including J. M. Wilson at Rugby School, who made the mistake of replying, and then had great trouble ending the correspondence.

Eventually, in desperation, Wilson tried sarcasm, writing

> Sir, I admire your indomitable perseverance, your hand-writing, and your liberal expenditure of postage stamps...

though the real 'killer' was

> your conclusions are as certain as their premises, with which they are in fact identical.

Sadly, this was all completely lost on James Smith, who went on to publish his ideas about π (Fig. 196), only to suffer very public ridicule.

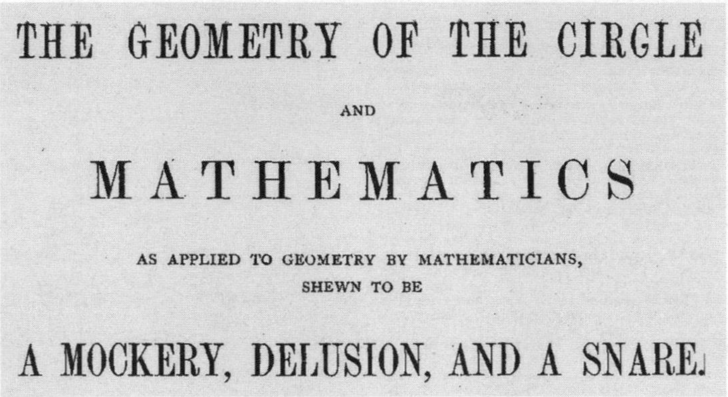

THE GEOMETRY OF THE CIRCLE

AND

MATHEMATICS

AS APPLIED TO GEOMETRY BY MATHEMATICIANS,
SHEWN TO BE

A MOCKERY, DELUSION, AND A SNARE.

Fig. 196 James Smith's *Geometry of the Circle* (1869).

The real trouble, for both Smith and Heisel, was that the Swiss mathematician Johann Lambert had proved as early as 1761 that π is irrational, so cannot be written exactly as the ratio of *any* two whole numbers, let alone 25/8 or 256/81.

Squaring the circle?

The irrationality of π is linked to the problem of 'squaring the circle', though the two are not quite the same thing.

That problem, which dates from ancient times, is to construct, in a finite number of steps, a square equal in area to that of a given circle, using only an unmarked ruler and compass.

And Ferdinand Lindemann proved in 1882 that this is impossible.

Yet 15 years later, Edwin J. Goodwin, a country doctor in Indiana, claimed to have done it, and persuaded his state representative, Taylor Record, to introduce a bill establishing his 'new mathematical truth' *as part of Indiana state law.*

THE MATHEMATICAL BILL.

Fun-Making In the Senate Yesterday Afternoon—Other Action.

The Senate yesterday afternoon occupied itself with House bills on second reading. The engrossment of a number of small bills was ordered, and the bill legalizing the incorporation of Poneta, Wells county, was passed under suspension of the rules.

Representative Record's mathematical bill legalizing a formula for squaring the circle, was brought up and made fun of. The Senators made bad puns about it, ridiculed it and laughed over it. The fun lasted half an hour. Senator Hubbell said that it was not meet for the Senate, which was costing the State $250 a day, to waste its time in such frivolity. He said that in reading the leading newspapers of Chicago and the East, he found that the Indiana State Legislature had laid itself open to ridicule by the action already taken on the bill. He thought consideration of such a proposition was not dignified or worthy of the Senate. He moved the indefinite postponement of the bill, and the motion was carried.

(a)   (b)

Fig. 197 (a) From *The Indianapolis News*, Sat. 13 Feb, 1897.
(b) Professor Clarence Waldo, whose intervention stopped the bill from becoming law.

The bill itself was extremely confused, with different parts of it seeming to imply no fewer than *six* different values for π, all of them rational.

And it was only when the bill reached the Indiana Senate that it came to grief, sparing further embarrassment, thanks to an intervention by a professor of mathematics at Purdue University, who happened to be visiting at the time.

Is every triangle isosceles?

Now and again, geometry is *deliberately* done wrongly, in order to make some cautionary point.

In Lewis Carroll's *Picture Book* of 1899, for instance, there is one of his favourite puzzles: a 'proof' that all triangles are isosceles (Fig. 198).

The 'proof'

Let the bisector of angle BAC meet the perpendicular bisector of BC at F.

Draw FH perpendicular to AB and FG perpendicular to AC.

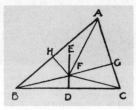

Fig. 198 Where's the mistake?

Triangles AFH and AFG are congruent (by angle-sum + ASA), so

  (i) AH = AG

  (ii) FH = FG

Triangles BDF and CDF are congruent (by SAS), so

  (iii) FB = FC.

Combining (ii), (iii), and Pythagoras' theorem gives

  (iv) HB = GC.

Adding (i) and (iv) gives AB = AC !

The mistake

If the angle-bisector of A meets the perpendicular bisector of BC at a single point F at all, that point always lies *outside* the triangle not inside (Fig. 199).

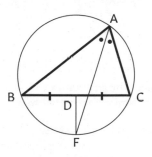

Fig. 199 The mistake.

It lies, in fact, on the circumcircle of ABC, midway between B and C, so that the arcs BF and FC, being equal, form the same angle at A, consistently with AF being the bisector of angle BAC.

In consequence of all this, when we drop perpendiculars from F onto AB and AC, one of the two lands on an *extension* of the side (AC in Fig. 199), rather than the side itself, and this invalidates the very last step in the 'proof'.

So, while some witty mathematician once said,

> Geometry is the art of correct reasoning on
> incorrect diagrams,

we are well advised, in general, to ensure that our diagrams are not *too* incorrect.

Malfatti's problem

In 1803, as part of a wider investigation, the Italian mathematician Gianfrancesco Malfatti considered the question

> Given a triangle of fixed size and shape, how do you
> construct 3 non-overlapping circles inside it so that their
> total area is as large as possible?

And he thought he knew the answer: choose the circles so that each one touches two sides of the triangle and both the other two circles (Fig. 200).

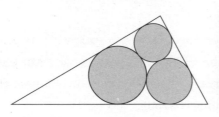

Fig. 200 Malfatti's problem.

Bizarrely, it was not until 1930—more than a hundred years after Malfatti's conjecture—that two mathematicians noticed something wrong *in the simplest possible case*—the equilateral triangle (Fig. 201).

For, in his configuration, the circles occupy a fraction 0.729 of the triangle's area, but we do very slightly better by first sticking in the biggest circle we can, and then two smaller ones.

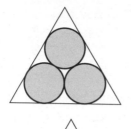

$$\frac{\pi\sqrt{3}}{(1+\sqrt{3})^2} = 0.729$$

$$\frac{11\,\pi}{27\sqrt{3}} = 0.739$$

Fig. 201 Something wrong with Malfatti's 'solution'...

Even more oddly—and 35 years later still—someone else noticed that we don't really have to do any calculations at all here to see that something is wrong.

For if the triangle in question is very long and thin, Malfatti's solution is rather obviously not correct, and we surely do better with the second arrangement in Fig. 202.

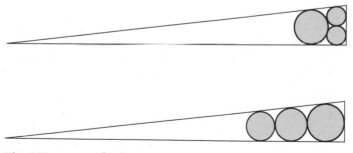

Fig. 202 ...something very wrong.

Finally, in 1967, someone showed that Malfatti's solution is *never* correct, no matter what the shape of the original triangle.

In geometry, then, a wrong conjecture can *remain* a wrong conjecture for a very long time.

New Angles on Geometry

For the past 200 years, classical geometry has continued to flourish, both through new results and through new ways of looking at old ones.

Dandelin spheres

In 1822, for instance, the Belgian engineer G. P. Dandelin discovered a novel way of relating the two views of an ellipse, namely as (a) a cross-section of a cone or (b) a curve for which PA + PB is constant (Fig. 203).

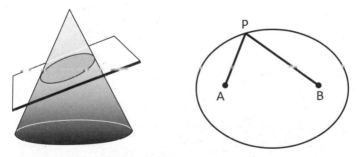

Fig. 203 Two views of an ellipse.

His idea was to introduce two *spheres* into the problem, one above the plane that slices through the cone, and one below (Fig. 204).

The sphere above the plane touches the cone along a horizontal circle (shown) and is just big enough to touch the plane at A. Similarly, the sphere below touches the cone along another horizontal circle (shown) and touches the plane at B.

Note that once we have made our particular slice, and got our particular closed curve, the two Dandelin spheres are fixed in position and size.

Our aim now is to prove that as a point P moves around the curve, PA + PB remains constant.

To do this, join P to the top of the cone by a straight line, and let that meet the two horizontal circles at A′ and B′.

Now comes the final, beautiful step...

As PA and PA′ are both tangent to the upper sphere, from the same external point P, they are *equal*. Similarly, PB and PB′ are equal, because both are tangent to the lower sphere.

So PA + PB = PA′ + PB′.

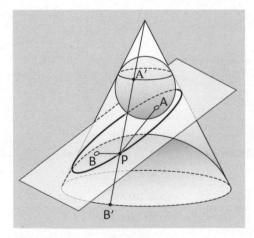

Fig. 204 Dandelin spheres.

But PA′ + PB′ *is constant*, because it is the fixed distance, measured straight down the side of the cone, between the two (fixed) horizontal circles.

So PA + PB is constant, too, and that completes the proof.

More unexpected meetings

Miquel's theorem

Take any old triangle ABC. Let A′ be any point on the side opposite A, and choose points B′ and C′ similarly.

Then the circles AB′C′, BC′A′, and CA′B′ *meet at a common point* (Fig. 205).

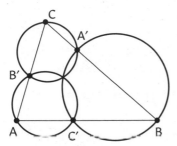

Fig. 205 Miquel's theorem.

This remarkable theorem, dating from 1838, is one of several discovered by the French school teacher Auguste Miquel.

And it has a most elegant proof (in Notes, p. 263) involving the four-points-on-a circle theorem of Chapter 11 *and* its converse!

Common chords

If three circles all intersect one another, the three common chords meet at a single point (Fig. 206).

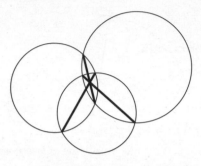

Fig. 206 Common chords.

This, too, has an elegant proof (Notes, p. 264) using three applications of the intersecting-chords theorem of Chapter 11.

I know little of the history of this particular result, but it appears as something of a throwaway remark on p. 337 of Hall and Stevens' *School Geometry* of 1904, so I imagine it is much older.

The seven-circles theorem

In 1974 an unusual book of just 68 pages appeared, with a preface which begins as follows:

> This collection of theorems in plane geometry, some of which we believe to be new, illustrates that there may still be plenty of interesting theorems awaiting discovery.

The main result involves a closed chain of six circles, each touching the next. And the whole chain touches the inside of a seventh circle (Fig. 207).

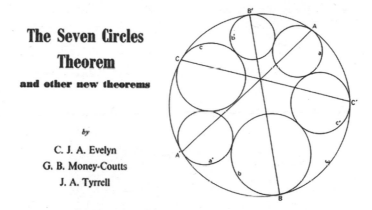

The Seven Circles Theorem

and other new theorems

by

C. J. A. Evelyn
G. B. Money-Coutts
J. A. Tyrrell

Fig. 207 The seven-circles theorem (1974).

Finally, if we join the points of tangency of 'opposite' circles in the chain, we find that the three lines meet at a single point.

In this kind of way, then, unexpected meetings in geometry continue to fascinate, even in modern times.

Non-periodic tiling

The year 1974 also saw the Golden Ratio

$$\phi = \frac{1+\sqrt{5}}{2}$$

make a remarkable reappearance in geometry through Roger Penrose's famous non-periodic tiling of the plane.

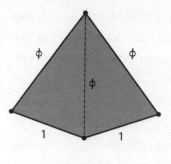
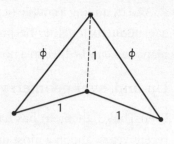

'KITE'   'DART'

Fig. 208 Penrose tiling.

In his most well-known example, just two types of tile are involved, the 'kites' and 'darts' of Fig. 208, both made from the triangles that occur in the regular pentagon on p. 59 and therefore involving the Golden Ratio.

And by adding a couple of simple rules about how the tiles can be fitted together he produced an example where the plane can *only* be tiled in a non-periodic (non-repeating) way.

Up-and-over geometry

Even Thales' theorem has had something of a makeover in recent years, and in a most unlikely context: the mechanism for operating a typical 'up and over' garage door (Fig. 209).

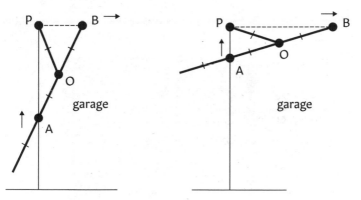

Fig. 209 Up-and-over geometry.

The door itself is divided into three equal parts by the points A and O.

The point A is constrained to move in a vertical shaft which runs up to the (fixed) point P at the top of the front of the garage. The point O, on the other hand, is attached to P by a rod with length one-third of the door height.

And a direct consequence of this arrangement, which has practical advantages, is that, when the door is opened, the top of the door B *moves exactly horizontally*.

And the reason for this is, essentially, Thales' theorem.

The only difference, really, is that one usually tends to think of Thales' theorem with AB as fixed and P as moving, whereas here the point P and the direction PA are fixed, and it is the diameter AB which moves!

And Finally...

Many years ago, someone tried to impress me with a triangle that had an angle-sum of 270°.

They took the surface of a sphere, and drew three lines along 'great circles', to produce the triangle in Fig. 210.

And it does, indeed, contain *three* right angles!

This is an example of *spherical geometry*, a branch of the subject with a long history and—not surprisingly—of great importance in navigation.

And it is in fact a special case of a beautiful theorem by

Fig. 210 Spherical geometry.

Girard (1629) which says that

$$\text{Angle-sum} = 180° \left(1 + \frac{4A}{S} \right),$$

where A is the area of the spherical triangle and S the surface area of the sphere.

So if the triangle is very small and almost flat, so that A is tiny compared to S, the angle-sum is almost 180°. For the triangle in Fig. 210, however, $A = \frac{1}{8}S$, and the angle-sum is, indeed, 270°!

Yet I'm sorry to say that, at the time, I was unimpressed.

'But that isn't really a *triangle*,' I said. 'Anyone can see that the sides aren't *straight*!'

I fear I lacked the imagination to see that they would certainly *seem* straight to anyone living—so to speak—in the purely two-dimensional world of the surface itself, unaware of anything 'outside'.

More seriously still, there was no one around at the time to tap me gently on the shoulder and say:

'Now, young man... you seem terribly certain that you know, in any conceivable context, what a "straight line" is. Just how would you *define* one, exactly?'

And, just possibly, if there *had* been someone around to say this, I would have had less trouble when I first encountered geometries in which some of our most familiar and deeply entrenched ideas just *don't hold*...

Non-Euclidean geometry

Throughout the history of geometry there have been numerous attempts to *prove* Euclid's parallel postulate from his other assumptions.

But all failed, and in the 1830s it finally became clear why: simply dropping Euclid's parallel postulate altogether, while keeping the others, can lead to quite different, but entirely self-consistent, geometries (Fig. 211).

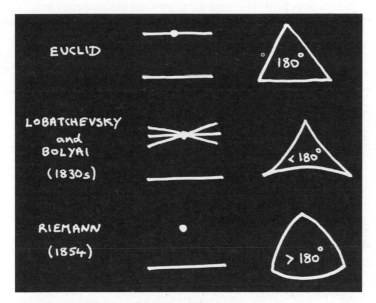

Fig. 211 Euclidean and non-Euclidean geometry.

This discovery is due essentially to the Hungarian mathematician Janos Bolyai and the Russian mathematician Nikolai

Lobatchevsky, though Bolyai had to overcome the strongest possible discouragement from his father:

> You must not attempt this approach to parallels. I know
> this way to the very end. I have traversed this bottomless
> night, which extinguished all light and joy of my life.
> I entreat you, leave the science of parallels alone.
>
> Letter to Janos Bolyai from his father, 4 April 1820

But Bolyai persisted, and we now know that there are self-consistent geometries with *infinitely many* lines through a given point parallel to a given line, and triangles with angle-sums which are always *less* than 180°.

And some years later, Riemann showed that there is a geometry, too, with *no* lines through a given point parallel to a given line, and triangles with angle-sums which are always greater than 180°.

A further weird feature of both geometries is that there are *no similar triangles that are not congruent*. This is because—as in the case of spherical geometry—the angle-sum depends on the size of the triangle.

Projective geometry

By the middle of the nineteenth century, the whole idea of geometry had opened up in other ways, too.

Projective geometry, for instance, came to prominence through problems of perspective in art, but something of its

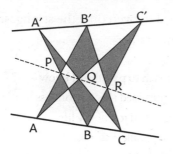

Fig. 212 Pappus' theorem.

spirit can be seen in a remarkable theorem due to Pappus of Alexandria, in about AD 320.

Let A, B, C be successive points on a line, and A', B', C' successive points on another (Fig. 212). Then let AB' and BA' meet at P, and define the points Q and R, similarly.

Then P, Q, and R *lie on a straight line*.

And the most distinctive aspect of all this is that there is no mention whatsoever of length, angle, or area.

Topology

This part of the subject is largely concerned with whether one geometrical object can (or cannot) be continuously deformed into another.

In Fig. 213, for instance, and despite appearances, the objects in (a) and (e) are *topologically* the same, because one can be gradually, and continuously, deformed into the other, through the sequence shown.

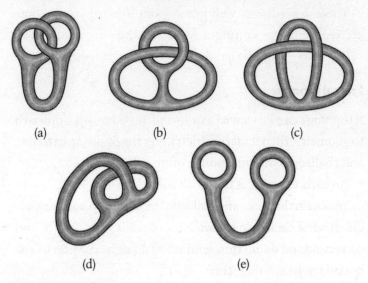

(a) (b) (c)

(d) (e)

Fig. 213 Topology.

The Möbius band in Fig. 214, however, is topologically different from an ordinary (untwisted) band, because it is impossible to deform one into the other without cutting or tearing it. And the one-sidedness of the Möbius band leads to some remarkable properties.

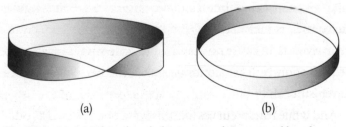

(a) (b)

Fig. 214 (a) A Möbius band. (b) A normal, untwisted band.

These are, I think, well known, but if you haven't come across them, try cutting a Möbius band—right along its centre-line—into two bands, and see what happens!

Fractal geometry

If topology can be viewed as a sort of 'broad brush' approach to geometry, then *fractal* geometry is at the opposite extreme, and challenges our intuition in quite different ways.

An early example is the 'Koch snowflake' of 1904.

To obtain this, take an equilateral triangle, replace the middle third of each side by *two* lines of equivalent length, and then continue doing this, with every straight portion of the boundary, *for ever* (Fig. 215).

Fig. 215 Making a Koch snowflake.

The result is a geometrical curve which encloses a finite area—just 8/5 times the area of the original triangle—yet the curve itself is infinitely long, and has no clearly defined tangent *anywhere*, because no matter how closely we look at some little portion of it we continue to see further detail; things never 'smooth out'.

And while fractal curves like this were once viewed as oddities, they now arise frequently in the subject of dynamics, where they are often linked with the idea of chaos.

Would you believe it?

In 1924, two Polish mathematicians, Stefan Banach and Alfred Tarski, proved that it is possible to dissect a solid sphere into a finite number of pieces, and then reassemble those pieces to form *two* solid spheres which are each *the same size as the original* (Fig. 216).

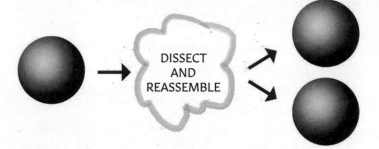

Fig. 216 The Banach–Tarski paradox.

It is true that the pieces in the 'in between' stage are so mathematically subtle that they have no clearly defined volume *at all*, but the result still stunned the mathematical world.

* * *

In short, there are whole new worlds of geometry out there, each with its own fresh surprises.

And in my experience, at least, the best surprises in geometry tend to stay with you, throughout life—a little bit like old friends.

In fact, to be quite honest, there are times when I wonder if this particular book has been gently brewing ever since my very first surprise in geometry, all those years ago, at school, one cold winter morning in 1956, when I was ten.

Notes

CHAPTER 6: PYTHAGORAS' THEOREM

The traditional source quoted in connection with the Chinese text *Zhou bi suan jing* (Fig. 33, p. 35) is Joseph Needham's *Science and Civilisation in China* (Cambridge, 1959), vol. 3, pp. 22–3.

But this is, I believe, acknowledged to be a very free translation, and a discussion of some of the difficulties can be found in Christopher Cullen's article 'Learning from Liu Hui? A Different Way to Do Mathematics' in *Notices of the American Mathematical Society*, vol. 49, pp. 783–90 (2002).

371 PROOFS OF PYTHAGORAS

Ann Condit's proof

In Fig. 217, ABC is our right-angled triangle, and P is the mid-point of the hypotenuse AB.

We begin by drawing (dotted) lines through P perpendicular to the sides, dividing all three squares in half.

The plan is to prove Pythagoras' theorem by proving that Area DQC + Area FRC = Area APG (and then multiplying by 4!).

By 'shearing' the first two triangles in a Euclid-like way this is equivalent to proving that

$$\text{Area DPC} + \text{Area FPC} = \text{Area APG},$$

which is the heart of Condit's proof.

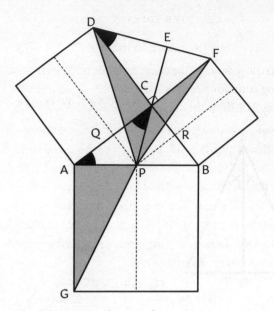

Fig. 217 Ann Condit's proof.

To do this, she joins DF and extends PC to meet DF at E.

Now, triangles ACB and DCF are congruent by SAS, and after a little angle-chasing, this leads to the important conclusion that PC is *perpendicular* to DF.

So, by viewing PC as the 'base' of the triangles DPC and FPC,

$$\text{Area DPC} + \text{Area FPC} = \tfrac{1}{2}\text{PC.(DE+EF)}$$

$$= \tfrac{1}{2}\text{PC.DF}$$

But PC = AP, and because of the congruence just noted, DF = AB = AG, and the result then follows.

CHAPTER 10: CONVERSELY...

Isosceles triangles

An alternative proof of the converse of the isosceles triangle theorem, avoiding contradiction, goes as follows.

In Fig. 218, draw the angle-bisector AD.

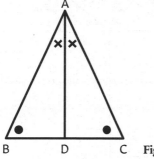

Fig. 218 Is it isosceles?

Then, as the angles of a triangle add up to 180°, $\angle ADB = \angle ADC$.

Triangles ADB and ADC are then congruent by ASA (with AD as the side in question), so AB = AC.

Euclid couldn't do this in *Elements* I.6, because he hadn't at that stage established either ASA congruence or the angle–sum of a triangle.

The converse of Thales' theorem

Yet another way of proving this is by just using Pythagoras' theorem—albeit somewhat relentlessly.

In Fig. 219, O is the mid-point of AB, with OA = OB = r, and $\angle APB = 90°$. And we want to prove that $R - r$, so that P lies on the circle with AB as diameter.

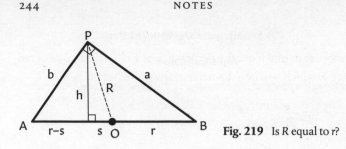

Fig. 219 Is R equal to r?

One application of Pythagoras gives $a^2 + b^2 = 4r^2$.
Two more turn this into

$$h^2 + (r+s)^2 + h^2 + (r-s)^2 = 4r^2,$$

which simplifies to $h^2 + s^2 = r^2$. One final application of Pythagoras gives $h^2 + s^2 = R^2$, so $R = r$.

CHAPTER 12: OFF AT A TANGENT

The slightly informal proof of the secant–tangent theorem (Fig. 67, p. 74) may be avoided, if desired, by appealing instead to the *alternate segment theorem*, which says that $\angle PTQ = \angle PRT$ in Fig. 220.

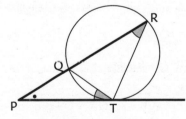

Fig. 220 Proving the secant–tangent theorem.

Triangles PTQ and PRT are then similar by AA, so

$$\frac{PT}{PR} = \frac{PQ}{PT},$$

whence the result.

The alternate segment theorem

To prove this, introduce the perpendicular to PT at the point of tangency T, which will be a diameter of the circle (Fig. 221).

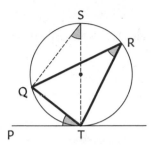

Fig. 221 Proving the alternate segment theorem.

Then ∠PTQ = ∠QST because of Thales' theorem, and ∠QST = ∠QRT because they are 'angles at the circumference' standing on the same chord QT.

Measuring the Earth

It is worth noting, perhaps, that we don't *have* to use the secant–tangent theorem in Fig. 68, p. 75. We could, instead, just use Pythagoras' theorem to give

$$(R+h)^2 = D^2 + R^2,$$

which leads directly to the same result

CHAPTER 13: FROM TANGENTS TO SUPERSONIC FLOW

Two additional results of great value in trigonometry arc

$$\sin(A+B) = \sin A \cos B + \cos A \sin B$$
$$\cos(A+B) = \cos A \cos B - \sin A \sin B,$$

and I offer a picture-proof in Fig. 222.

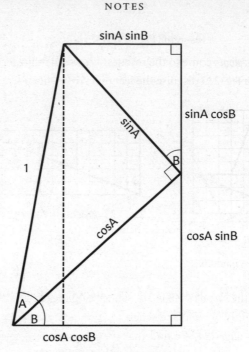

Fig. 222 Finding $\sin(A+B)$ and $\cos(A+B)$.

CHAPTER 14: WHAT IS π, EXACTLY?

Two important special cases of the results in the Note for Chapter 13 are the *double-angle formulae*:

$$\sin 2A = 2 \sin A \cos A$$
$$\cos 2A = \cos^2 A - \sin^2 A.$$

In view of Pythagoras' theorem $(\cos^2 A + \sin^2 A = 1)$, the second of these is equivalent to

$$\cos 2A = 2 \cos^2 A - 1.$$

Setting $A = \theta/2$ then leads to the 'useful little result from trigonometry' which I used, at the very end of Chapter 14, when deriving Viète's remarkable formula for π.

Inspector Euclid Investigates...

Mercifully, *slopes* come to the rescue, because the thick lines on the diagram in Fig. 223 disguise the fact that ABC is not a straight line.

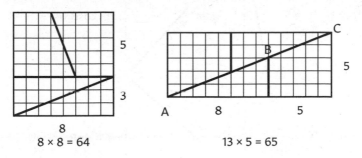

Fig. 223 A paradox...

Plainly, the slope of AB is 3/8, while the slope of BC is 2/5, which is very slightly greater, and when we draw the second figure more accurately (Fig. 224) we find that the re-assembled pieces leave a very long, thin, parallelogram in the middle. This gap accounts exactly for the apparent discrepancy.

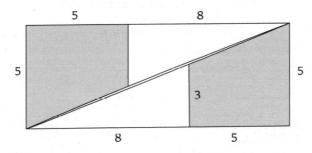

Fig. 224 ...and its resolution.

This puzzle is often attributed to the great Sam Loyd, who is said to have presented it at the American Chess Congress in 1858. But puzzles of this general kind have a very long history, as can be seen from David Singmaster's article 'Vanishing Area Puzzles', in *Recreational Mathematics Magazine*, no. 1 pp. 10–21 (2014).

CHAPTER 17: GEOMETRY AND CALCULUS

The slope of the curve $y = x^2$

To find the slope at the point P, in Fig. 225, choose a nearby point Q, also on the curve.

If the x-coordinates of P and Q are x and $x + h$, say, then their corresponding y coordinates will be x^2 and $(x + h)^2$.

In consequence, as we move from P to Q,

$$\frac{\text{increase in } y}{\text{increase in } x} = \frac{2xh + h^2}{h},$$

and on cancelling the factors of h we have

$$\frac{\text{increase in } y}{\text{increase in } x} = 2x + h.$$

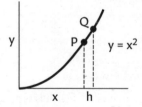

Fig. 225 Finding the slope of the curve $y = x^2$.

Finally, we notice that the closer we take Q to P, i.e. the smaller we take h, the closer this ratio comes to $2x$, so that the slope of the curve itself *at* P must be $2x$.

The parabola: focus and directrix

Associated with the parabola $y = x^2$ there is a special point called the *focus* F, with coordinates $(0, \frac{1}{4})$, and a special (horizontal) line called the *directrix* with equation $y = -\frac{1}{4}$.

A key property is that any point P on the parabola, with coordinates (x, y), is the same distance from F as it is from the directrix, so that

$$PF = PD$$

in Fig. 226.

To prove this, the distance formula in Fig. 97, p. 104, gives $PF^2 = x^2 + (y - \frac{1}{4})^2$ and $PD^2 = (y + \frac{1}{4})^2$, and these turn out to be equal, because $y = x^2$.

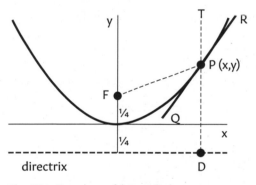

Fig. 226 Geometry of the parabola.

The parabola: reflection property

Now let the line QPR bisect $\angle$FPD in Fig. 226. As PF = PD, FPD is an isosceles triangle, so this angle-bisector will be perpendicular to the triangle's 'base' FD.

As F has coordinates $(0, \frac{1}{4})$ and D has coordinates $(x, -\frac{1}{4})$, the slope of FD is $-\frac{1}{2x}$.

The slope of QPR must therefore be 2x, which, as we have seen, is *the slope of the parabola itself* at P.

So QPR must be the *tangent* to the parabola at P.

Moreover, as ∠QPD and ∠TPR are opposite angles, it follows that

$$\angle FPQ = \angle TPR,$$

which is the reflection property.

The area under a curve

To understand the claim made in Fig. 104, p. 112, imagine *x* increasing very slightly, as in Fig. 227. The area A will also increase slightly, and the increase will almost correspond to a tall thin rectangle of height *y*.

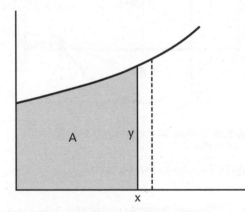

Fig. 227 A small increase in area.

In other words, the small increase in area will *almost* be obtained by multiplying *y* by the small increase in *x*.

Or to put it another way, if we divide the small increase in A by the small increase in *x* we will get, very nearly, *y*.

And that is, essentially, why the slope of the curve in Fig. 104b, p. 112, is y.

CHAPTER 18: A ROYAL ROAD TO GEOMETRY?

To establish Malton's 'extraordinary property of the Circle' in Fig. 110, p. 120, draw diameters AC and BD in Fig, 228 to create a rectangle ABCD.

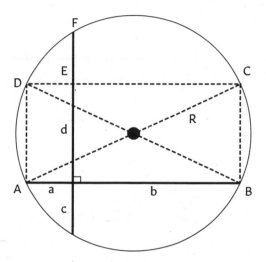

Fig. 228 Proving Malton's 'extraordinary property'.

Then appeal to symmetry to claim that EF = c, and therefore BC = $d - c$.

Finally, use Pythagoras' theorem on triangle ABC to obtain

$$(a+b)^2 + (d-c)^2 = (2R)^2.$$

And on multiplying out, the result emerges, because *any* two intersecting chords have $ab = cd$, whether they are at right angles or not (Fig. 64, p. 72).

CHAPTER 19: UNEXPECTED MEETINGS

The perpendicular bisectors

In Fig. 229, let D, E, and F be the mid-points of the sides, and let the perpendiculars from D and E meet at O.

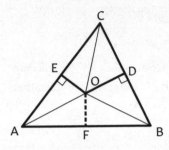

Fig. 229 Proving that the perpendicular bisectors are concurrent.

Triangles OEA and OEC are congruent by SAS, so OA = OC.

Similarly, OC = OB.

Therefore OA = OB.

Now join OF.

Triangles OFA and OFB are then congruent by SSS.

So ∠OFA = ∠OFB, and OF is therefore the *perpendicular* bisector of AB.

The angle-bisectors

In Fig. 230, let the angle-bisectors from B and C meet at I. Draw ID, IE, IF perpendicular to the sides.

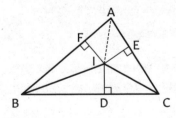

Fig. 230 Proving that the angle-bisectors are concurrent.

Then triangles IDB and IFB are congruent (by angle-sum + ASA). So IF = ID.

Similarly, IE = ID.

So IF = IE, and triangles IFA and IEA are therefore congruent (by Pythagoras + SSS).

So ∠IAF = ∠IAE, and IA is therefore the *bisector* of angle BAC.

The Euler line

In Fig. 231, consider one perpendicular bisector OC′, one median CC′, and extend OG by twice its length to a point that we will rather presumptuously call H, so that GH = 2 × OG.

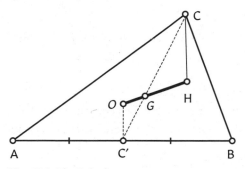

Fig. 231 The Euler line.

As GC = 2 × GC′, the triangles OGC′ and HGC are then similar by SAS, so ∠HCG = ∠OC′G . This means that CH is parallel to OC′, and therefore perpendicular to AB.

In other words, the point we have tentatively called H *lies on the altitude through* C.

By repeating the argument, starting with a different corner and opposite side, H lies on *all three* altitudes, and must therefore, indeed, be the orthocentre.

The pedal triangle

In Fig. 232, the four points A, F, H, and E all lie on a circle (with AH as diameter, in fact), by the converse of Thales' theorem (twice). So the angles *a* are equal, as they both stand on the same chord EH (Theorem (b) in Fig. 60, p. 68).

Similarly, the angles *b* are equal.

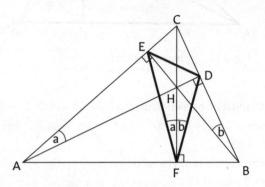

Fig. 232 The pedal triangle.

Finally the angles ∠CAD and ∠CBE are equal, because they are both 90° − ∠ACB.

So *a* = *b*, and the altitude CF therefore bisects ∠EFD. The same argument then applies to the other two altitudes.

The nine-point circle

In Fig. 233, we use the suffix 1 to denote the feet of the altitudes, 2 for the mid-points of the sides, and 3 for the mid-points of the lines joining H to the corners of the triangle.

By the mid-point theorem, C_2B_2 is parallel to BC.

Again by the mid-point theorem, B_2C_3 is parallel to AH, and therefore *perpendicular* to BC.

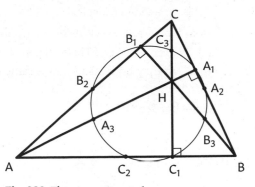

Fig. 233 The nine-point circle.

So B_2C_3 must be perpendicular to C_2B_2, i.e. $\angle C_2B_2C_3 = 90°$.

Similarly, $\angle C_2A_2C_3 = 90°$, and we already know that $\angle C_2C_1C_3 = 90°$.

So, by the converse of Thales' theorem (three times!) we learn that the points B_2, A_2, and C_1, all lie on a circle with C_2C_3 as diameter.

To put it another way, the (unique) circle which passes through the mid-points of the sides, A_2, B_2, and C_2, *also* passes through the points C_1 and C_3.

The same argument (with different starting-points) then shows that it must also pass through B_1 and B_3 *and* A_1 and A_3, establishing the nine-point circle.

CHAPTER 20: CEVA'S THEOREM

Ceva's theorem is in the same spirit as a much earlier one, due to Menelaus of Alexandria, in about AD 100.

Menelaus' theorem

This concerns a triangle ABC which has its sides cut by a straight line, as in Fig. 234.

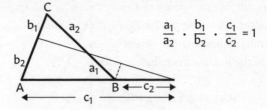

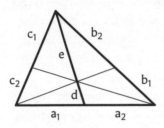

Fig. 234 Menelaus' theorem.

The simplest proof I know involves drawing the dashed line though B parallel to AC. Then, by one pair of similar triangles, its length is $\frac{b_1 a_1}{a_2}$, and by another, $\frac{b_2 c_2}{c_1}$, from which the result follows.

Ceva's theorem; another proof

Ceva's theorem can, in fact, be *proved* from Menelaus' theorem.

Fig. 235 Proving Ceva's theorem.

With reference to Fig. 235, two applications of Menelaus' theorem give

$$\frac{d}{e} \cdot \frac{c_1}{c_2} \cdot \frac{(a_1 + a_2)}{a_2} = 1$$

and

$$\frac{d}{e} \cdot \frac{b_2}{b_1} \cdot \frac{(a_1 + a_2)}{a_1} = 1,$$

from which the result follows.

And, as our original proof of Ceva's theorem was by means of area (Chapter 20), while we have just proved Menelaus' theorem using similar triangles, the above proof provides yet another example of the link between the two ideas.

CHAPTER 21: A KIND OF SYMMETRY

Two results from trigonometry

Trigonometry can be especially useful when dealing with a general triangle that has no particular symmetry or simplifying features.

In particular, two major results are

$$\text{Area} = \tfrac{1}{2}ab\sin C$$

and the *cosine rule*

$$c^2 = a^2 + b^2 - 2ab\cos C$$

(see Fig. 236a).

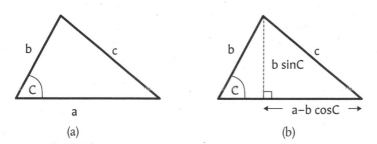

(a) (b)

Fig. 236 Triangle trigonometry.

The first of these is really just '$\tfrac{1}{2}$ base × height ', while the second comes from applying Pythagoras' theorem to the right-angled triangle in Fig. 236b with hypotenuse c.

Heron's theorem for the area

This can be established—after considerable algebra—by eliminating the angle C between the results above, using Pythagoras' theorem, i.e. $\sin^2 C + \cos^2 C = 1$.

Incircle and circumcircle

Figure 237 gives a 'picture proof' for the radius of the incircle, by simply viewing r as the height of the three constituent triangles, and adding the areas.

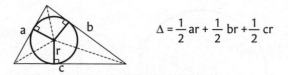

$$\Delta = \frac{1}{2}\,ar + \frac{1}{2}\,br + \frac{1}{2}\,cr$$

Fig. 237 Finding the radius of the incircle.

Figure 238 gives a picture proof for the radius of the circumcircle. It uses the main circle theorem—'the angle at the centre is twice that at the circumference'.

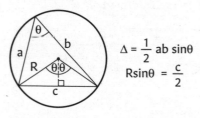

$$\Delta = \frac{1}{2}\,ab\,\sin\theta$$
$$R\sin\theta = \frac{c}{2}$$

Fig. 238 Finding the radius of the circumcircle.

Newton and the altitudes

This delightfully down-to-Earth proof that the altitudes meet at a point can be found in *The Mathematical Papers of Isaac Newton*, vol. 4,

p. 454 (Cambridge University Press, 2008), along with a substantial historical footnote by the editor, D. T. Whiteside.

The eyeball theorem

In Fig. 239, triangles OQB and OTO′ are similar, so QB/TO′ = OB/OO′.

If we let $D = OO'$ this gives

$$QB = \frac{rr'}{D},$$

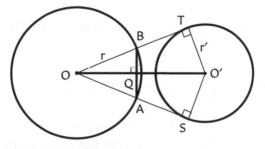

Fig. 239 The eyeball theorem.

The medians, by coordinate geometry

The formulae for the coordinates of a general point P on a line (Fig. 132, p. 143) follow from *similar triangles* (Fig. 240), for then $\lambda = (x - x_1)/(x_2 - x_1)$, with a similar result for the y-coordinates.

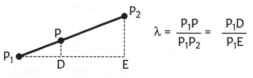

$$\lambda = \frac{P_1P}{P_1P_2} = \frac{P_1D}{P_1E}$$

Fig. 240 A point on a line.

CHAPTER 24: A SOAP SOLUTION

Much more about the geometry of soap films can be found in *The Science of Soap Films and Soap Bubbles* by C. Isenberg (Dover Publications, 1992).

CHAPTER 25: GEOMETRY IN *THE LADIES' DIARY*

A penny-farthing problem

Pythagoras applied to a certain right-angled triangle gives

$$(a+b)^2 = (a-b)^2 + D^2,$$

whence the result (Fig. 241).

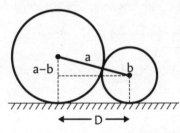

Fig. 241 Penny-farthing problem.

A tricky problem

The first solution to this problem appeared in 1731 and can be found on p. 336 of vol. 1 of Hutton's *The Diarian Miscellany* (1775). It makes an appeal to a 'curious theorem' by Ozanam, and entirely misses the neat, symmetric form of the final answer.

A second solution appears on p. 187 of vol. 1 of Thomas Leybourn's *The Mathematical Questions Proposed in the Ladies' Diary* (1817). The neat, symmetric form of the answer is given, together with a generalization of the original problem, but it appeals to some sophisticated theorems from advanced trigonometry.

I have managed to construct a solution involving only elementary trigonometry. It begins by using some Pythagoras and similar triangles to establish the results in Fig. 242.

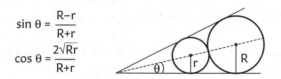

$$\sin\theta = \frac{R-r}{R+r}$$

$$\cos\theta = \frac{2\sqrt{Rr}}{R+r}$$

Fig. 242 Nested circles.

The three θ-like angles in the actual problem—say α, β, γ—have the property that $\alpha + \beta + \gamma = 90°$ (because the angles of a triangle add up to 180°), and I use this to eliminate them. The most effective way of doing this that I could see using elementary trigonometry is through

$$\cos(\alpha+\beta) = \cos\alpha\cos\beta - \sin\alpha\sin\beta = \sin\gamma,$$

but even that leads to some fairly tedious algebra before a quadratic equation for R emerges, and the neat, symmetric form of the answer:

$$R = \sqrt{ab} + \sqrt{bc} + \sqrt{ca}.$$

While the whole approach is strictly elementary, it is rather scrappy, and I cannot help thinking that there must be a better way.

CHAPTER 26: WHAT EUCLID DID

SSS congruence

Euclid's proof of SSS congruence, Prop. I.8, is based on the idea that SSS is enough to define a triangle uniquely, proved in Prop. I.7, and the proof of that is based on Fig 243, where he assumes (for contradiction) that the 'third' point could be in two different places, C and C′.

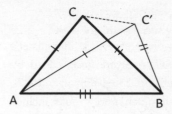

Fig. 243 Euclid's proof of SSS congruence.

There are then two *isosceles triangles* present. So, to paraphrase Euclid slightly:

$$\angle ACC' = \angle AC'C < \angle BC'C = \angle BCC',$$

which implies that $\angle ACC' < \angle BCC'$, which is absurd.

ASA congruence

His proof of ASA congruence is again by contradiction, and in much the same spirit as his proof of the converse of the isosceles triangle theorem in Fig. 55, p. 63.

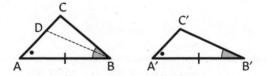

Fig. 244 Euclid's proof of ASA congruence.

The essence is as follows. Suppose the ASA conditions allow two different triangles, so that, for instance, AC > A'C' in Fig. 244. Then choose the point D so that AD = A'C', and draw BD.

The triangles DAB and C'A'B' will then be congruent by SAS, so $\angle ABD = \angle A'B'C'$. So $\angle ABD = \angle ABC$, which is absurd.

CHAPTER 30: WHEN GEOMETRY GOES WRONG...

For more about the 1897 episode in Indiana, see David Singmaster's article 'The Legal Values of Pi' in *Pi: A Source Book*, edited by L. Berggren, J. Borwein, and P. Borwein, pp. 236–9 (Springer, 1997).

That book includes, not surprisingly, much more about π, including (i) an extensive chronology (pp. 288–305), (ii) Lambert's 1761 paper proving irrationality (pp. 129–40), and (iii) Lindemann's paper proving that π is transcendental (pp. 194–206).

CHAPTER 31: NEW ANGLES ON GEOMETRY

Miquel's theorem

In Fig. 245, let two of the circles meet at P.

By the circle theorem shown in Fig. 63 (p. 71) we have $\angle A'PB' = 180° - c$ and $\angle B'PC' = 180° - a$, from which it follows that $\angle A'PC' = a + c$.

But from triangle ABC we have $a + c + b = 180°$, so that $\angle A'PC' + b = 180°$. It then follows, from the *converse* of the circle theorem just quoted, that the four points A'PC'B lie on a circle.

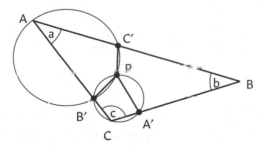

Fig. 245 Proving Miquel's theorem.

Common chords

In Fig. 246, let two of the common chords, AB and CD, meet at P.

Extend EP to meet circle 1 at F′ and circle 3 at F″.

Now apply the intersecting chords theorem (Fig. 64, p. 72) *three times*, to circles 1, 2, and 3 *in turn*:

$$EP.PF' = CP.PD = AP.PB = EP.PF''.$$

So PF′ = PF″, and therefore both F′ and F″ must, in fact, be the point F.

So P lies on the chord EF.

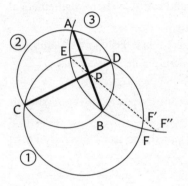

Fig. 246 Common chords.

Further Reading

GENERAL

The Penguin Dictionary of Curious and Interesting Geometry, by David Wells (Penguin, 1991). (A true miscellany, and great for dipping into on a cold winter's evening.)

A Square Peg in a Round Hole; Adventures in the Mathematics of Area, by Chris Pritchard (Mathematical Association, 2016).

Icons of Mathematics, by C. Alsina and R. B. Nelsen (Mathematical Association of America, 2011).

The Changing Shape of Geometry, ed. Chris Pritchard (Cambridge University Press, 2003). (Remarkable collection of essays.)

The Parsimonious Universe, by S. Hildebrandt and A. Tromba (Copernicus, 1996). (Beautifully illustrated account of maximum and minimum problems.)

The Pythagorean Theorem, by Eli Maor (Princeton University Press, 2007).

THE HISTORY OF GEOMETRY

Geometry by its History, by A. Ostermann and G. Wanner (Springer, 2012).

Mathematical Thought from Ancient to Modern Times, by Morris Kline (Oxford University Press, 1972).

EUCLID'S ELEMENTS

The classic edition, originally dating from 1903, is:

The Thirteen Books of Euclid's Elements, trans. Sir Thomas Heath (Dover, 1956).

It includes a great deal of scholarly commentary by Heath.

A much more recent, and relatively compact edition, also based on Heath's translation, was published by Green Lion Press in 2002.

A striking, full-colour facsimile of Oliver Byrne's extraordinary 1847 colour edition of the first six books of Euclid's *Elements* was published by Taschen in 2010.

In addition, there is a much admired website version by David Joyce, at https://mathcs.clarku.edu/~djoyce/elements/elements.html

I have also found the following commentary on the *Elements* very helpful:

Euclid—The Creation of Mathematics, by B. Artmann (Springer, 1999).

TEXTBOOKS OF GEOMETRY

When I was first learning the subject, many years ago, my principal books were *Teach Yourself Geometry*, by. P. Abbott (English Universities Press, 1948) and *Elementary Geometry*, by C. V. Durell (Bell, 1925).

The second part of Abbott's book is especially concise, and an excellent modern textbook in a similar spirit, but with many valuable exercises, is *Crossing the Bridge*, by Gerry Leversha (UK Mathematics Trust, 2008). A more expansive treatment can be found in *Geometry*, by Harold Jacobs (Freeman, 2003).

At a more advanced level—and especially if you enjoyed the 'unexpected meetings' of Chapter 19—I recommend:

The Geometry of the Triangle, by Gerry Leversha (UK Mathematics Trust, 2013) as well as Clark Kimberling's famous website *Encyclopedia of Triangle Centers* at http://Faculty.evansville.edu/ck6/encyclopedia/ETC.html.

For looking ahead further still, towards university-level differential geometry, I recommend:

Elementary Geometry, by John Roe (Oxford University Press, 1993).

THE TEACHING OF GEOMETRY

Teaching and Learning Geometry, by Doug French (Continuum, 2004).

Mathematics for the Multitude? A History of the Mathematical Association, by Michael Price (The Mathematical Association, 1994).

The History of the Geometry Curriculum in the United States, by Nathalie Sinclair (Information Age Publishing, 2008).

AND FINALLY...

Five *very* different ways of looking far beyond the geometry in most of this book:

Things to Make and Do in the Fourth Dimension, by Matt Parker (Particular Books, 2014).

Gems of Geometry, by John Barnes (Springer, 2009).

Mathematics: The Science of Patterns, by Keith Devlin (Scientific American Library, 1994).

Topology: A Very Short Introduction, by Richard Earl (Oxford University Press, 2019).

Ideas of Space, by Jeremy Gray (Clarendon Press, Oxford, 1989).

Acknowledgements

I wish to thank Richard Earl, Gerry Leversha, and Vicky Neale for some very helpful comments on the manuscript. I am also extremely grateful to Latha Menon, Jenny Nugee, and Emma Slaughter of Oxford University Press for taking such great care at all stages of production.

Publisher's Acknowledgements

Epigraph for Chapter 12:

Twelve (12) words from *Decline and Fall* by Evelyn Waugh (Penguin Books 1937, 1962) Text Copyright © Evelyn Waugh, 1928

From *Decline and Fall* by Evelyn Waugh, copyright © 2012. Reprinted by permission of Little, Brown and Company, an imprint of Hachette Book Group, Inc.

Picture Credits

2.	ScienceCartoonsPlus.com
11.	AF Fotografie / Alamy Stock Photo
12.	The Bodleian Libraries, The University of Oxford, MS.D'Orville 301, Fols.21v-22r
20.	INTERFOTO / Alamy Stock Photo
25.	E.F. Smith Collection, Kislak Center for Special Collections, Rare Books and Manuscripts, University of Pennsylvania
26b.	Yale Babylonian Collection, YBC 7289
27.	Granger Historical Picture Archive / Alamy Stock Photo
33.	akg-images / Pictures From History
34.	Felix Bennett
43.	Northcliffe Collection/ANL/Shutterstock
72.	Z J Levensteins, Naval Ordnance Laboratory
85 (left).	GL Archive / Alamy Stock Photo
89.	From van Schooten, *Exercitationum Mathematicorum* (1657)
93.	Cambridge University Library
102a.	GL Archive / Alamy Stock Photo
107 (right).	The Picture Art Collection / Alamy Stock Photo
129.	Sir Isaac Newton, Geometria curvilinea and Fluxions, MS Add. 3963, p 54r Reproduced by kind permission of the Syndics of Cambridge University Library
134.	Image by George Rex
142 (right).	Pictorial Press Ltd / Alamy Stock Photo
143 (right).	Pictorial Press Ltd / Alamy Stock Photo

144.	Public Domain
145 (left).	History and Art Collection / Alamy Stock Photo
147.	Public Domain
181.	From the author's private collection
182a.	Lebrecht Music & Arts / Alamy Stock Photo
182b.	Photo: Blackwell's Rare Books
189.	From James M. Wilson, An Autobiography, Sidgwick and Jackson, 1932
190.	The Mathematical Association
191.	Public Domain
195.	From the author's private collection
196.	Public Domain
197b.	Public Domain
200 (right).	The History Collection / Alamy Stock Photo
p. 12 (top)	19th era / Alamy Stock Photo
p. 43 (top, right)	Copyrighted 1938. Associated Press. 2164996:0820PF
p. 43 (bottom, left)	Wellesley College Archives, Library & Technology Services
p. 136	From William Jones, 'A New Introduction to the Mathematics', London, 1706
p. 137 (top)	FALKENSTEINFOTO / Alamy Stock Photo
p. 178 (bottom)	History and Art Collection / Alamy Stock Photo

The publisher and author apologize for any errors or omissions in the above list. If contacted they will be pleased to rectify these at the earliest opportunity.

Index

A

AA similarity 55, 59, 67, 72, 244, 256, 259
Al-Biruni 75
alternate angles 6, 90
alternate segment theorem 245
altitudes 122, 132, 140
ancient Greece 18
angles 5
  alternate 6
  at the centre 68
  at the circumference 68
  bisected 125, 127, 252, 254
  corresponding 4
  in straight line 5
  measurement 5, 24
  notation for 16
  obtuse 38
  on same arc 68
  opposite 6
  right 6
  sum of triangle 7, 119, 193, 203, 204, 210
Archimedes 18, 86, 134, 215
area 24
  of circle 86
  of rectangle 25
  of similar triangles 47
  of triangle 25, 37, 87, 99, 192, 257
  preserved 38, 41, 98, 195, 241
  using calculus 112, 250
area-ratio theorem 130
ASA congruence 51, 54, 218, 223, 253, 262
Association for the Improvement of Geometric Teaching 206
Aubrey, J. 36

B

Babington, J. 27
Banach-Tarski paradox 238
Barrow, Isaac 12, 108
Benson, L. S. 208
bisector
  of angle 125, 252
  perpendicular 123, 252
Bolyai, J. 233
Byrne, O. 178

C

calculus 108, 248
cannonballs 91
Carroll, Lewis 198, 218
Cavalieri, B. 156
centre of gravity 134
centroid 125, 253

Ceva's theorem 129
  converse 131
  from Menelaus' theorem 256
Chinnery, G. 94
chord 16
circle 16
  angles on same arc 68, 178, 219, 245, 254
  area of 86
  four points on a 71, 225, 263
  intersecting chords 72, 251, 264
  main theorem 68, 77, 92, 258
  nine-point 127, 254
  packing 89
  tangent to 73
  terminology 16
  theorems 68
circumcentre 123. 127, 253
circumcircle 123, 139, 258
circumference 16, 86, 172
common chords theorem 226
concurrence:
  altitudes 122, 124
  angle bisectors 125, 252
  circles 225, 263
  chords 226
  Gergonne point 133
  medians 125, 142
  perpendicular bisectors 123, 252
Condit, A. 43, 241
cone 94
congruence 14, 52, 54
  describing 54
  see also ASA, SAS, SSS

conic section 94
constructions 161
contradiction, see proof: by contradiction
converse 61
  of Ceva's theorem 131
  of circle theorem 161, 225
  of isosceles triangle theorem 62, 243
  of Pythagoras's theorem 61
  of Thales' theorem 65, 153, 243
Cooley, W. D. 209
coordinates 101
corresponding angles 4, 23, 190
cosine rule 257
$\cos \theta$ 80
cyclic quadrilateral 71

D

Dandelin spheres 223
Descartes, R. 101
diagonal of rectangle 48, 64
Dido's problem 150
differentiation 110
directrix 249
distance by coordinates 103
Dodgson, C. 198, 207
double-angle formulae 246

E

Earth, measurement of 22, 75
ellipse 94, 223
  in Fermat problem 155
  and planets 97
  reflection property 96

equilateral triangle 16, 81, 90, 157, 161, 162, 220, 237
equilibria, multiple 168
Einstein, Albert 46, 175
Eratosthenes 22
Euclid 9, 76, 114
Euclid's *Elements* 9, 12, 36, 39, 180
  ASA congruence 185, 262
  Barrow's edition 12
  Book I 85, 192
  criticism 208
  Oliver Byrne's edition 178
  postulates 182
  SAS congruence 185, 186
  similarity and area 194
  SSS congruence 185, 261
  structure 180
  teaching of 206
Eudoxus 181
Euler, L. 127, 137, 253
eyeball theorem 141, 259

F

Fermat, P. 101, 154
focal points of ellipse 95
focus of parabola 110, 249
fractal 237

G

Galileo 84
garage door geometry 229
Gauss, C. F. 123
Gergonne point 133
Girard 232
golden ratio 58, 227

Goodwin, E. J. 217
gradient, *see* slope
gravitation 85, 99
Gutierrez, A 141

H

half-angle formula 93
Heisel, C. T. 213
Heron of Alexandria 52
  area formula 138, 258
  reflection of light 52
hexagon 88, 90, 198
Hobbes, T. 36, 113
Hofmann, J. E. 163
hyperbola 94

I

icosahedron 182
incentre 125
incircle 125, 139, 258
inclined plane 84
infinite product 92
infinite series 136
Inspector Euclid investigates… 106, 247
integration 112
intersecting chords theorem 71
irrational number 28, 137, 181, 214, 216
isosceles triangle 14, 17, 62, 69, 187, 218, 243

K

Kepler, J. 91, 97
Koch snowflake 237

L

ladder problems 27, 45
Ladies' Diary, The 171
   keeping one's head 171
   penny-farthing problem 174, 260
   tricky problem 176, 260
   window problem 173
Lambert, J. 137, 216, 263
Lecchi, A. 197
Leibniz, G. L. 110, 136
Levett, R. 206
Lindemann, F. 216, 263
line:
   parallel 4
   straight 5, 10, 232
Lobatchevsky, N. 233
Loomis, E. S. 42
Ludlam, W. 200

M

Mach number 80, 82
Mahavira 45
Malfatti's problem 220
Malton, T. 114, 120
Mankiewicz, Johanna 175
Mathematical Association,
   The 206, 212
maximisation 111, 147, 150, 220
mechanics in geometry 134, 164
medians 125, 132, 142, 259
Menelaus' theorem 255
midpoint theorem 56
minimisation 110, 147, 149, 154, 165
Miquel's theorem 225, 263

mirror image 15, 53
Möbius band 236

N

Newton, Sir Isaac 38, 98, 110,
   140, 258
nine-point circle 127, 254
non-Euclidean geometry 233

O

octahedron 182
opposite angles 6, 11, 184
   of cyclic quadrilateral 71
orthocentre 124, 253

P

packing problems 89
Pappus 188, 235
parabola 94, 101, 110, 249
parallel lines 4, 189, 201, 202
   and alternate angles 6, 190
   and corresponding angles 4,
   190
   distance apart 201
   'same direction' 202
parallel postulate 184, 189, 233
   alternatives 199
parallelogram 54
pedal triangle 127, 254
penny-farthing problem 174, 260
Penrose, R. 227
pentagon 58, 182
perpendicular 5
   bisectors 123, 252

perspective 234
Philo of Byzantium 186
pi:
 definition 86
 exact formula for 92
 infinite product 92
 infinite series 136
 irrational 137, 216, 263
 link with imaginary numbers 137
 symbol for 136
 value of 89, 137, 215
pizza theorem 121
planetary motion 97
Plato 19
Platonic solids 182
Playfair's postulate 199
point 10
 Fermat 155
 Gergonne 133
polygon, regular 87, 159
polyhedron 182
'practical work' 23, 210
projective geometry 234
proof:
 by contradiction 62, 208
 by rotation 162
 false 203, 204, 218
 importance of 2
Pyramid, Great 20
Pythagoras' theorem 26
 3-4-5 special case 27, 60, 117
 Ann Condit's proof 43, 241
 converse 61
 Euclid's proof 39
 in China 34, 241

 in converse of Thales'
  theorem 243
 in coordinate geometry 103
 in *The Ladies' Diary* 172, 260
 in trigonometry 80, 257
 measuring the Earth 225
 proof 31, 32, 34, 36, 46, 196, 241

Q

quadrilateral 57, 70

R

radius 16
Recorde, R. 116
rectangle 25, 40, 48, 64, 107, 173,
  196, 247, 250, 251
reductio ad absurdum, see proof: by
  contradiction
reflection of light 52, 96, 110, 147,
  160, 249
Regiomontanus 76
regular polygon 87, 159
Riemann, B. 233
right angle 6, 183
road networks 164
rotation 162
Royal Road to Geometry 114, 120, 231
rugby 77
ruler and compasses 216

S

SAS:
 congruence 14, 41, 51, 53, 63, 185,
  187, 188, 192, 218, 242, 252, 262
 similarity 55, 56, 253

scale factor 55
secant-tangent theorem 74, 244
semicircle 1, 16, 65, 104, 150
seven circles theorem 226
Sherlock Holmes 44
shortest network 165
similar triangles 20, 45, 55
    and golden ratio 59
    describing 54
    link with area 48
    right-angled 20, 44, 46, 48
    Thales and 20
    see also AA, SAS, SSS
Simpson, T. 145, 152, 160, 184,
        193, 201
Simson, R. 183
sin θ 80
slope:
    of a curve 109, 248, 250
    of straight line 102, 247
    perpendicular lines 102
Smith, J. 215
snowflake 237
soap films 164, 167, 260
sphere 88
    packing 91
spherical geometry 231
square 25, 30
SSS:
    congruence 51, 62, 186, 252,
        253, 261
    similarity 55
Steiner, J. 153
squaring the circle 216

superposition 185
supersonic flow 79, 82
symmetry 15, 64, 123, 125, 138, 143,
        175, 177, 231, 261

T

tangent 73, 79, 110
    secant-tangent theorem 74
tetrahedron 182
Thales' theorem 1, 14, 16, 63, 230
    and Galileo 84
    by coordinate geometry 104
    converse 65, 153, 243, 254, 255
theorem:
    alternate segment 245
    angle bisector 125, 252
    Ceva's 129
    circle 68, 71, 72
    common chords 226, 264
    eyeball 141, 259
    Girard's 232
    intersecting chords 71
    isosceles triangle 14
    Menelaus' 255
    Miquel's 225, 263
    midpoint 56
    Pappus's 235
    Pythagoras', see Pythagoras's
        theorem
    secant-tangent 74, 244
    seven circles 227
    Thales', see Thales' theorem
    Varignon's 57
    Viviani's 157, 159

Thue, A. 91
tiling 227
topology 235
Torricelli, E.
    Fermat problem 155
    trumpet 113
triangle:
    altitudes 122, 140
    angle-bisectors 125
    angle sum of 7, 119, 193, 203,
        204, 210
    area of 25
    equilateral 16
    incentre 125
    isosceles 14
    medians 125, 142
    obtuse 38
    perpendicular bisectors 123
    right-angled 25, 26
    similar 20, 44
trigonometry 80
    addition formulae 245, 261
    and altitudes 132
    and pi 93
    and soap films 166
    and supersonic flow 83
    area of triangle 257

cosine rule 257
double-angle formulae 246

u

up-and-over geometry 229

v

Varignon's theorem 57, 148
Viète, F. 92, 101
viewing angle 76
Viviani, V. 156
volume:
    sphere 88
    trumpet 113

w

Waldo, C. 217
Ward, J. 33, 117, 202, 209
Whitehead, A. N. 209
Wilson, J. M. 205, 208, 215

Y

Young Mathematician's Guide
        33, 117

Z

Zhou bi 35

1089 AND ALL THAT

A Journey into Mathematics

David Acheson

978-0-19-959002-5 | Paperback | £8.99

'Every so often an author presents scientific ideas in a new way…Not a page passes without at least one intriguing insight…Anyone who is baffled by mathematics should buy it. My enthusiasm for it knows no bounds.' *Ian Stewart, New Scientist*

'An instant classic…an inspiring little masterpiece.' *Mathematical Association of America*

'Truly inspiring, and a great read.' *Mathematics Teaching*

This extraordinary little book makes mathematics accessible to everyone. From very simple beginnings Acheson takes us on a journey to some deep mathematical ideas. On the way, via Kepler and Newton, he explains what calculus really means, gives a brief history of pi, and introduces us to chaos theory and imaginary numbers. Every short chapter is packed with puzzles and illustrated by world famous cartoonists, making this is one of the most readable and imaginative books on mathematics ever written.